DIMENSIONS OF GENDER PROBLEMS
POLICIES AND PROSPECTS

DIMENSIONS OF GENDER PROBLEMS

POLICIES AND PROSPECTS

Edited by

Deepak Bishoyi

DISCOVERY PUBLISHING HOUSE
NEW DELHI-110002

First Published - 2007

Reprinted - 2018

ISBN: 978-81-8356-192-1

Dimensions of Gender Problems
Policies and Prospects

Published by:

DISCOVERY PUBLISHING HOUSE PVT. LTD.
4383/4B, Ansari Road, Darya Ganj
New Delhi-110 002 (India)
Phone: +91-11-23279245, 43596064-65
Fax: +91-11-23253475
E-mail: discoverypublishinghouse@gmail.com
sales@discoverypublishinggroup.com
web: www.discoverypublishinggroup.com

Printed at:
Infinity Imaging Systems
Delhi

Foreword

As part of a research study a couple of years ago, I was eliciting answers to a range of questions in formal interview settings amongst members of rural communities in the Southern Indian States. I had insisted upon doing the interviews myself because it had been quite a few years since I had engaged myself in data collection. Indeed, it was a great learning experience. Reading and teaching various themes and process of social change, I suppose, I may have begun to take things for granted. Certain patterns in the manner in which responses were being given were eye openers to me on certain persisting rural realities. Among the many patterns one that is of relevance to the present context pertains to the question of gender preferences. To my question on the number of children a couple had, I was beginning to get the response that was totally misleading. Nearly 70 per cent of the respondents—men and women—had been answering by merely indicating the number of male children. It was as though the female children were not to be 'counted'. After a few interviews, I had to specifically ask separately for the number of Male and Female children. Two decades ago, this had been the pattern. I must admit that over the years I had begun to expect that the girl child's existence (if not status) had come to be recognised ('counted'). Perhaps I was being too naïve, ignoring the forces of gender preference in my own society.

Yet, decades of scholarship, research and debates, have not produced the intellectual climate that has taken gender beyond rhetorics. As part of the course work oriented to the doctoral students in the Institute for Social and Economic Change (Bangalore), the students are taught papers on gender studies and research methods. A batch of young students, ably supported by former teachers and friends elsewhere decided to bring together a few literature survey-based essays related to the theme of gender studies. Like many other colleagues in the Institute, I was too pleased upon learning of their efforts.

I am happy that these essays are now brought out in a book form and M/s Discovery Publishing House has agreed to publish it. While congratulating the contributors led by Mr. Deepak Bishoyi, I do hope their work will receive the critical acclaim it deserves. I am hopeful that the essays here will benefit both the students and practitioners in the field of development with a keen sensitivity to gender issues.

G. K. KARANTH

Professor of Sociology

Head, Centre for Study of Social Change and Development

Institute for Social and Economic Change

Nagarabhavi

Bangalore–560 072

Preface

Gender refers to the socio-cultural definition of man and woman; the way societies distinguish men and women and assign them social roles. The distinction between sex and gender was introduced to deal with the general tendency to attribute women's subordination to their anatomy. For ages it was believed that the different characteristics, roles and status accorded to women and men in society, are determined by biology (i.e., sex), that they are natural, and therefore not changeable. For centuries, women have been victims of social prejudices and discrimination. Lakhs of female foeticides are the evidence that even today, parents prefer male child. Women are generally presumed to be weak, passive, dependent and people oriented. On the other hand, men are considered strong aggressive, independent and things-oriented. Assumptions become reality when society prepares males and females for performance in presumed roles. Unequal treatment to women is an important issue of human rights and social justice. Women represent half the resources and half the potential in all the societies. Efforts to promote greater equality between women and men can also contribute to the overall development of the society. The empowerment, autonomy of women and the improvement of women's social, economic and political status are essential for the achievement of sustainable development in all areas of life.

The National Commission for Women at the Center and State Commission for Women at the State level have been constituted under the Statutes and ample legal provisions have been legislated providing penal action for violations in a bid to improve the conditions of women, in general, has remained unaffected by all those Statutes and Rules. The mind-set of the implementers of the

Statutes and Rules remains unchanged and so is the position of women. Family considers women as a social responsibility, an economic liability, not an asset anywhere. The economic contribution of the women to her family is named as performing the household chorus. The probable reason may be that the members of the family are satisfied that the women have to perform those as rituals in return to the family for the security she is provided. She is not treated equally with her male counterparts. Her performances are never computed in terms of financial contributions, but only social obligations for proper discharge of which she is very rarely thanked, but for any irregularity she is invariably made accountable. The women have a peculiar position in the family. Even girl child born to a family is considered an unwelcome guest. Till her marriage she is treated as a security of some other family. The family members are deeply concerned about her marriage. They prefer to save money to meet her marriage requirements than invest for her personality development. During her entire life span, she remains a living commodity. After marriage, she become a newcomer member and apprentice for her new family keeping. She is praised as long as she learns the lessons of self-sacrifice and condemned the moment she tries to assert her individuality. She can only be visualized and accepted as a docile daughter- in –law, faithful wife and loving mother, but can not be recognized as a women individual. She is exposed to all the undesirable circumstances the moment there is any deviation or any shift of attitude and becomes an unnecessary commodity at even sometimes by near relatives. Rape or sexual allurements often make her unwed mother bringing with it all the companion harassments and hardships. The thought of any such possibility keeps haunting parents and is kept under protectionism, which severely shrinks her freedom and space of manoeuvrability for development.

The above are only a small portion of a long chain of misery confronting women of all age groups in the present day society. Women are constantly living under pressure of the incidents like distorted personality development due to discrimination, threat of or actual occurrences of deception, rape, kidnapping, dowry tortures, murder, etc. In considering the situation faces by women in the society we took a step to gather some scholarly articles, which focused, to the problem.

As a part of doctoral fellow course work in ISEC, we have to deal with several gender-related issues that give birth to several articles by me and some of my other colleague scholars. We thought that these must reach to the people who are in academia, administration and other developmental sectors who are interested in this field. In this regard we are thankful to Dr. K.G. Gayathri Devi for providing us congenial environment to grow our ideas. I am thankful also to other eminent personnel and research scholars who agreed to write scholarly articles for the present book. So, this book is an outcome of conglomeration of different ideas from academicians and research scholars.

The study is of paramount importance in understanding and formulating policy related to various dimensions focused to gender problems. It is hoped that researchers of social science, women studies, developmental studies, population studies and general readers interested in the problem would appreciate the work.

There are 14 selected papers relating to the various issues on gender problem. The paper contributors have made all efforts to present their ideas and the editor personally expressed his deep sense of gratitude for their contributions, which have enriched this volume.

My highest regards are acknowledged to my family members especially my father whose blessing, moral support and sound advice are responsible for my progress in this study and throughout my academic career. My heartly regards are to Dr. S.N. Tripathy, Reader in Economics, Aska Science College, Aska, Mr. M.N. Pattanayak, Lecturer in Economics, Kandupadar College, and Mr. K.C. Pradhan, Lecturer in Economics, Khemundi College, Digapahandi without their blessings and moral support, I may not be able to carryout the present work. I am grateful to all the friends in ISEC, Bangalore and IIPS, Mumbai for providing me all moral support to finalize this work. I am also thankful to Mr. K.S. Narayana of ISEC, for doing painstaking task of copy editing, which was quite demanding.

Last, but not the least, is the commendable support provided by Sri Tilak Wasan, Proprietor and other associated personnel of Discovery Publishing House, New Delhi, for having agreed to publish this volume and giving it the shape in which it is placed for scrutiny of the academia, administrators and others interested in the field.

Deepak Bishoyi
Editor
Institute for Social and Economic Change (ISEC)
Bangalore

Contents

List of Contributors

Mr. Akshya Panigrahi is Ph.D. Scholar, Institute for Social and Economic Change (ISEC), Nagarbhavi, Bangalore, 560 072.

Mr. Ananta Basudev Sahu is Research Officer, International Institute for Population Sciences, Govandi Station Road, Deonar, Mumbai, 400 088.

Mr. Avinandan Taron is Ph.D. Scholar, Institute for Social and Economic Change (ISEC), Nagarbhavi, Bangalore, 560 072.

Mr. Bibhu Prasad Sahu is Research Scholar, Department of Economics, Berhampur University, Orissa, 760 007.

Mr. Chandre Sekhar Bahinipati is Ph.D. Scholar, Madras Institute of Development Studies (MIDS), Chennai, 600 020.

Mr. Deepak Bishoyi is Ph.D. Scholar, Institute for Social and Economic Change (ISEC), Nagarbhavi, Bangalore, 560 072.

Dr. Dillip Kumar Mishra is Lecturer in Economics, Nayagarh (Auto) College, Nayagarh, Orissa, 752 069.

Dr. L. Ladusingh is Professor and Head, Department of Mathematical Demography and Statistics, International Institute for Population Sciences, Govandi Station Road, Deonar, Mumbai, 400 088.

Mr. Manas Ranjan Pradhan is Ph.D. Scholar, International Institute for Population Sciences, Govandi Station Road, Deonar, Mumbai, 400 088.

Miss. Sailaja Nandani is Ph.D. Scholar, Institute for Social and Economic Change (ISEC), Nagarbhavi, Bangalore, 560 072.

Miss. Sancheeta Ghosh is Ph.D. Scholar, Institute for Social and Economic Change (ISEC), Nagarbhavi, Bangalore, 560 072.

Mr. Sangram Kishor Patel is Ph.D. Scholar, International Institute for Population Sciences, Govandi Station Road, Deonar, Mumbai, 400 088.

Mr. Sontosh Kumar Sahu is Research Scholar, Department of Economics, Berhampur University, Orissa, 760 007.

Dr. Usha Ram is Reader, Department of Public Health and Mortality Studies, International Institute for Population Sciences, Govandi Station Road, Deonar, Mumbai, 400 088.

1

Gender Issues on Socio-economic Development

Mr. Deepak Bishoyi

Women constitute almost half of the population and the major number of work hour, i.e. 65 per cent is undertaken by women. However, the return she gets in comparison with man is very meagre; she gets only 10 per cent of the world's income. In the developing countries there is discrimination often found in the payment of wages. Women are paid much less than men for the same work for example, construction workers, agricultural labourers, factory workers, etc. The United Nation declared 1975-1985 as the United Nations Decade for Women, mainly to achieve gender equality. Gender has been defined as 'socially determined differences in roles and responsibilities of women and men'. Gender equality is being talked importantly very often nationally and globally in order to ensure social justice. It is one of the aspects of human development, the remaining being social and economic equality.

The social construct of gender is beginning to crumble in face of the onslaught of liberalization. This implies the liberalization of thought process in order to empower individuals, communities and nations to seek their own destiny. Many commentators on gender observe with varying degree of seriousness that women have plenty of influence but not much authority. Women, they say, are the power behind the throne, the stage managers who work behind the throne, and the stage managers who make it all happen and then retreat to the wings, leaving the men to take the curtain calls. According to this line of reasoning female power is based on womanly wiles and seductive persuasion as powerful as strength. Since the women get exactly what they want why make things even unfair by giving them legitimate authority too.

During the middle and the late 19th century, however, reality began to dawn at least for the daughters of the middle and upper classes. As the industrial revolution separated home and workplace, it became a status symbol for man to 'provide' for his family through his career, while women confined to domestic roles in the home. Although some historians have seen this development as an occasion for women to carve out a new sphere of power over family and home increasing immersion in the domestic sphere could also become a trap, severely affecting their capacity to act freely or pursue economic activities.

It was at this time that same middle and upper class women joined together in voluntary associations such as missionary societies or women's clubs to show their concern over child labour, juvenile delinquency, alcohol abuse, and factory safety conditions. They joined the first generation of women college graduates who were inspired to seek careers and transform the world they found around them. By the early 20th century women suffrage had become a primary objective of such groups.

The vote for women was seen not only as a significant step towards equal legal status, but also a step towards achieving social reform, cleaning up government and politics. The suffrage, it belived was a mortal sin to leave the home and take a job were now as a matter of patriotic necessity to help win the war by replacing a soldier gone to the front. In India we had the Freedom Movement that made every woman belonging to the affluent class come out of the four walls and join the freedom struggle. Millions of middle class women too discovered that, notwithstanding what culture had taught them, they were fully capable of running their own lives and playing role in the work force as well as the home.

The major economic reason for a specific gender focus in poverty programmes is that the poorer a family the gender their reliance on the earning capacity of women. Even with the greater their reliance on the earning capacity of women. Even with the most effective economic development policies, most poor families would not be able to survive without a continued major economic contribution from their female members. This is becoming recognized in China and can seen to influence the debate on urban employment policies for women.

However, women typically earn lower wages and have much more limited access than man to productive resources such as land, credit, fertilizer and higher technology, with the consequence their productive capacity, and their ability to contributing to the survival or improvement of their families is severely constrained. Despite the accumulating evidence about the crucial economic role of women in low income families, government often think in terms of women needing 'income generating activities' to perform in their spare time as extension of domestic activities in contrast to men needing 'employment' to provide for the needs of their families. Understanding and removing the constraints of productive female employment must therefore be a major component of all poverty elimination strategies.

A number of studies have shown that not only do women have less access than men to the benefits of economic modernisation, but it is also not uncommon for the conditions of significant number of poor women getting worse as a result of an economic reform programmes which were introduced to benefit the poor.

Education, work participation, exposure to mass media, access to resources and political participation are important means by which women gain autonomy and empowerment. The extent of gender inequalities in each of these domains reflect women's position in the society.

Gender Inequalities in Education: The key role of education in individual functional development is well recognised. Women's education is known to accrue greater benefits to the family. Large educational differences between men and women tend to sustain and perpetuate gender inequality within the family and society at large. Gender gap in education is a major determinant of low status of women in India.

In patriarchal societies, investment in boy's education is considered worthwhile because it is an investment in their future contribution to the family wealth. As a corollary the higher participation of young females in household works is reflected in their lower school participation rates, and schooling per se is considered more a liability than an asset in the case of girls. A more educated groom may demand a high dowry.

South Asia is one of the regions where the level of gender gap is higher. The progress in reducing the gender in literacy and education has been very slow in India, due to strong son preference. The overall gender gap in literacy shows an increasing trend until 1991. For the first time, there was a marginal decline in gender gap in literacy during 1991-2001.

Gender and Employment: The major source of women's economic worth is deprived from their participation in the paid labour force. In fact, female employment has a basis for women's independence and worthy of emphasis in both the development and feminist literature (Blumberg, 1991). Female labour force participation is associated with the greater economic independence and access to money. It affects three components of women's position: extent of exposure to the outside world, social and economic interactions and the level of autonomy in decision making within and outside the family.

Female labour force participation is known to have the following possible benefits:

(a) Improves the return to investment in girls;

(b) Improves the status of women in society, and therefore the value attached to the young girls;

(c) Lowers the dowry levels and therefore reduces cost of rearing the daughters;

(d) Makes women less dependent on adult sons for security in old age, and therefore, reduces son preferences; and

(e) Improves the bargaining power of the adult women and their ability to resist male pressure to discriminate against girls.

In contrast, in traditional patriarchal societies, women do not typically participate in economic activities outside home, but take responsibility—for a large share of secondary agricultural activities (Bishoyi, 2005).

Strong cultural norms effectively discourage women from seeking work that is outside the home (Arokiasamy, 2002). Men are assigned with primary responsibilities for activities involving the market place or that occur outside the home in which women have prime responsibility for the management of both children and the home. Major investment decisions (such as those concerning the human capital of children and the acquisition of sales of assets) are considered men's responsibilities. Such cultural norms have been important barriers to female participation in work outside their home.

There are two hypotheses: one that gender discrimination in motivation by cultural factors i.e. son preferences and the other that scarcity of resources may heighten the gender bias, Bhasin (2000) suggested neglect of female children is related to low female labour force participation. On the other hand, Miller (1981) has started that the exclusion of women from production is further linked to exclusion from holding property and access to resources.

Female Labour Participation and Occupation Pattern

The demand for labour and distribution of wealth are two determinations of female labour force participation. The demand for labour depends upon the agriculture intensity of irrigation, and presence or absence of large labour supply. Several studies have recognised female labour force participation as a useful measure of women's status and in an important correlate of north-south sex ratio differentials in India (Miller, 1981, Murthi et. al 1991,). They have argued that the greater female autonomy in the south largely stems form higher female labour force participation rate due to widely prevalent wet-rice cultivation. Consequently, not only the labour force participation is high in the southern states but also the share of employment outside their farm business is the highest.

Unpaid work involving women in a major way is still substantially underestimated in national statistics on the labour force, GNP and national income. Four sectors predominate in the under estimation of unpaid work of female labour-subsistence production, the household economy, the informal sector, and volunteer work. Domestic production and related activities where women are largely engaged in have been totally excluded as it is perceived simply to fall outside the conventional definition of work. The exclusion of all these sectors in national income mostly affects the contribution of women's work. The exclusion implies that the influence of this production is not accounted in market price determination and economic production. The international labour organization is engaged in conceptualisation and development of methodology of data gathering on this aspect of labour force participation.

Gender Differentials in Access to Resources or Ownership of Assets: The practice concerning property ownership and inheritance rights varies by religious groups. The mainstream Hindu traditional practice is to deny daughters rights on parents property and to give some limited rights in the husband's property, except in smaller communities where women have fairly generous rights e.g. in south India, northeast India and among some tribal communities.

Despite the enactment of gender-just property laws, this is yet to be accepted as a social practice. Denial of property rights strongly inhibits women's capacity to make more effective contributions besides lessening decision making power in economic matters.

Political Representation of Women: Political representation of women in India both in notional parliament and state assemblies has been marginal and small. While the ratio of males to females contesting in national elections still remains 30 times higher in favour of male, the percentage winning among female candidates contesting election is higher for females than males. Longterm trend in the thirteen Lok Sabha elections indicates very little improvement in terms of female representation. Panchayat Raj Institutions has a 33.3 per cent representation of women in accordance with the 73rd constitution amendment bill. However still it remains to be implemented across the states.

Household Headship: In all cultures, where patriarchal values still in exist varying degrees, almost universally men head the households, hold major decision making authority. In India, only 10 per cent of the households are female headed. A major portion of this female headship is due to the death of husband.

Economic Empowerment

Although women experienced an employment boom in the late 1960s and 1970s the work in most cases, offered no possibilities for economic advancement and upward mobility. Economic empowerment is that conscious state of womanhood where she would find herself equipped to 'outside' practical world. This need leads to an increase in the number of women workers from the late 40s to the present decade. All during 1980s and 1990s women's employment rate increased four times faster than that of men.

But an important point one has to remember is that when women went out to work, it was not to prove their equality with men, rather women were taking jobs in order to help the family a traditional 'female role'. They were underpaid, denied promotion opportunities and treated for the most part as 'marginal' workers. Nevertheless, their employment proved crucial to the economy and society. More and more of them came from better educated middle class families. Their employment provided millions of families to improve their economic condition. Otherwise, providing a college education for one's children would have been virtually impossible.

Social Empowerment

However, despite constitutional guarantee and penal enactments, women are treated inhumanly in most part of the century. They suffer from gender inequality. Throughout their lives, women suffer gender discrimination, in the womb, at home, and at the work place. The practice of female foeticide, female infanticide, and preferential treatment to boys for food and health care has been obvious in most of the societies of the world. At 933 girls for 1000 boys, India's sex ratio portrays the household inequality, economic and political inequality, social brutality and educational inequality.

According to the Child Marriage Restraint Act (the Sharads Act) the minimum age for a girl to marry is 18 years. But in practice, it has been violated in many states. According to UN Population Fund Report 2000, 4 per cent of women in India are subject to various types of domestic violence. As recently as on 8 March 2002, a bill was introduced in Lok Sabha on the Protection from Domestic Violence Bill 2002, which seeks to provide that if any relative of the victim subjects her to habitual assault or makes her life miserable or injures or harms her, would constitute Domestic Violence. Similarly the Anti Dowry Act has not acted as a severe deterrent to prevent dowry deaths across the country. Giving and taking huge amounts of dowry is widely accepted in middle and upper class families. Money tempts the husband, mother-in law, to grab more and more from the bride's family. In case of negative response, brides are harassed, beaten, tortured and burnt. The National Crime Records Bureau recorded 4935 cases of dowry figure shot up to 6232 in the year 2000 (Das, 2002).

The government should examine seriously the possibility of reopening some of these cases wherein the ends of justice have not been met and explore to what extent those who had connived in the process would not continue to remain beyond the ambit of the law. In addition to some non governmental organizations' crusade against this burning issue, statutory bodies like the State Women Commission should not be found wanting in serving as pressure groups and sentinels on the palpable laxity in the enforcement of the well intended piece of legislation.

Political Empowerment

The issues faced by the Indian women elected to the Panchayats and other public offices are not unlike those that historically were faced and sometimes are still faced by Western Women making their voices heard through political process that was denied to them not so long ago. There is a parallel in the development of Women finding a political voice in India and the League of Women Voters in USA founded in 1920, where women got the right to vote. But prior to that, that whole period building up to women getting suffrage, one of the issues women had was involved with the Temperance Movement trying to get men where so many women at the panchayat level got elected on that platform. The National Commission for Women (NCW). Established by Government of India in 1992 seems to be working on a non-partisan fashion. They try to transcend the party problems and the party situation to look at domestic violence and violence against women at election time. The NCW members are doing a lot of investigative work; there is a tremendous difference between sitting in a national organization and just talking about the issues, and going into the communities and seeing first hand what goes on, and they are doing just that and caring about what happens.

Lack of money is another thing that Indian women who are engaged in political process face. Even though women haven't been in full force until the last 10 years or so. And even then, women did not feel free to spend their money in political activities or send it to a political candidate. They may give it to a charity or hospital or university or something like that, but to send money to a political candidate is almost unheard of.

Women constitute about 49per cent of the total population. But at the state and national level their political empowerment is negligible. Between 1999 and 2000, only 220 women MLAs were found in states and UTs. Had the Women's Reservation Bill been enacted the total number of women MLAs would have gone up to 1480. The poor representation of women in Lok Sabha is seen from the fact that the number of women MPs in the Parliament has never crossed 50 at any time in the past, out of 760 member in both the Houses.

Constitutional Empowerment

Democratic countries in the world have incorporated several provisions in their constitution to ensure gender equality and dignity of women. It would be worthwhile at this juncture to look at the constitutional guarantees and special enactments offered to the women in India. First, the rights guaranteed to the women by the constitution of India.

1. Equality before law;
2. Prohibition of state discrimination based on religion, caste, sex, place of birth;

3. Special provision for women and children;
4. Equality of opportunity in public employment;
5. Equal pay for equal work;
6. Provision for maternity relief;
7. Reservation of seats for women in panchayats and municipal bodies.

In spite of all these constitutional guarantees and amendments there still does exist dichotomy between the preached and the practised. The advancement and development of India in 21st century requires equal participation of women to prove that they too are as competent as men in any field.

A multi pronged effort needs to be made as a relatively long term strategy so that even the poorest of the poor women become active participants in the country—both in government and private sectors. This would involve the formidable task of educating all women in all the intricacies of our polity. Moreover, some kind of support structures needs to be provided. For instance, the drudgery of housework has to be removed by making adequate arrangements for childcare and domestic work. A culture of sharing household responsibilities among men and women would need to be created so that the cultural edifice need to be created so that the cultural edifice of non-productive work as an exclusive domain of women gets dismantled.

Finally, various efforts need to be made to be seen that women remain above all these and do not get influenced for any reason. They will have to function with great objectivity, keeping uppermost in mind the overall interest of economic and social development of areas where they live. A philosophy like this should not be difficult to implement if women integrate themselves fully in the multifarious activities of our national life.

REFERENCES

Arokiasamy, P. (2002), "Gender Preference, Contraceptive Use and Fertility in India: Regional and Development Influences", *International Journal of Population Geography*, Vol. 8, 49-67.

Bhasin (2000), *Understanding Gender*, Kali Primaries, New Delhi.

Bishoyi, D. (2005), "IT—*New Mantra for Women's Empowerment*", Social Welfare, August, Vol. 52, No. 5.

Blrmberg, R.L. (1991), "Introduction: The Triple Overlap of Gender Stratification, Economy and the Family", in: *Gender, Family and Economy: the Triple Overlap*, ed. R.L. Blrmberg, Newbury Park: Sage Publication, pp. 7-34.

Miller, B.D. (1981), *The Endangered Sex*. Ithaca: Cornell University Press, Cornell.

Murthi, et al. (1991), "Mortality, Fertility and Gender Bias in India: A District Level Analysis", *Population and Development Review*, 21 (4): pp. 745-782.

Pattanik, S. and Pradhan, T.M. (2005), "Empowerment of Women by Ending Gender Violence", Social Welfare, August, Vol. 52, No. 5.

Sen, G. et al. (1994), *Population Policies Reconsidered: Health and Empowerment and Rights*, Harvard University Press, Harvard.

World Bank (1991): Gender and Poverty in India—A World Bank Country Study, Washington, D.C.

2

Gender and Economic Development

Mr. Avinandan Taron

People often use the "gender" as a synonym for "sex". Sex however refers to the characteristics that make someone male or female. Gender has also been misused as a synonym for women or female. Development projects catering to the needs of women are often mistakenly referred to as gender needs. Instead *gender* refers to the socio-cultural construction of rules and relationships between male and women. In describing socio-cultural construction, gender analysis considers other social structures such as race, ethnicity class and caste.

Gender roles and relationships are the assigned activities and relative position in society of men and women. These help to determine the access to resources and opportunities based on local cultural perceptions of masculinity and feminity. While gender roles and relationships impose expectations and certain limitations on both men and women, they can perpetuate forms of subordination. Gender roles frequently mirror images leading some people to suggest that the division of male and female gender roles is inherent and set in stone. Gender roles are learned and can change over time. These roles are constructed through forces such as culture, tradition, politics and need varying from culture to culture and often from one social group to another within the same culture (according to characteristics such as class, ethnicity, race, age, caste and marital status). Recognizing that relationships are gendered, allows for the issue of power to be addressed.

Gender discrimination remains pervasive in different forms of life worldwide. This is so despite considerable advances in gender equality in recent decades. The nature and extent of discrimination vary considerably across countries and regions, but the patterns are striking. In no region of the developing world are women equal to men in legal, social and economic rights. Gender gaps are widespread in access to and control over resources, in economic opportunities, in power and political voice. Women and girls bear the largest and most direct costs of these inequalities, but the costs cut more broadly across society, ultimately harming everyone. For these reasons, gender equality is a core issue which has attained the importance of developmental objective in its own right. The rationale behind incorporating gender into development is to strengthen a country's ability to grow, reduce poverty and, at the same time, govern efficiently. Promoting gender equality should be such that it would be a part of development strategy that seeks to improve the standard of living of all people, men and women alike.

It is a general consensus that economic development opens many avenues for the reduction of gender inequality, but again it should be remembered that the growth will not alone disburse its effects. Institutional developments will also have to take place simultaneously to support such changes in the direction that provides for the equal opportunities for both men and women. Policy back up is an important factor to eradicate persistent inequalities. Although policies are important the benefits accrue only when the monitoring of such policy implementations are done.

The last half of the twentieth century saw great improvement in the absolute status of the women and in gender equality in most of the developing countries. The major contributors to such development are education and health. With a few exceptions, female education has increased considerably. The primary school enrollment rates has about doubled in South Asia, Sub Saharan Africa, Middle East and North Africa and rising even faster than that of the boy's. This has reduced the large gender gap in schooling. Women's life expectancy has increased by 15-20 years in the developing countries. More women have joined the labour force since 1970 with the participation rising by 15 average points in the East Asia and Latin America. This growth was larger than that of men, thus narrowing the gender gap in employment. *Despite the progress significant gender inequalities in rights, resources and voice persist in all developing countries and in many areas the progress has been slow and uneven. Moreover, socio-economic shocks in some countries have brought setbacks jeopardizing hard-won gains.*

With regards to *rights* women share less rights than men in terms of social, political, economic and legal rights. In a number of countries women still lack independent rights to own land, manage property or conduct business. Gender disparities in rights constrain the sets of choices available to women in many aspects of life, often profoundly limiting their ability to participate in or benefit from development. Women continue to have systematically poorer command over a range of productive *resources,* including education, land, information and financial resources. Despite recent increases in women's educational attainment, women continue to earn less than men in the labour market even when they have the same education and years of experience as men. Women are often restricted to certain occupations in the developing countries. Limited access to resources and weaker ability to generate income whether in self-employed activities or in the wage employment constraint women's power to influence resource allocation and investment decisions in the home. Unequal rights and poor socio-economic status relative to men also limit their ability to influence decisions in their communities as well as in the national level. Women remain vastly underrepresented in national and local assemblies, accounting for less than 10 per cent of the seats in the parliament, on an average (except in East Asia where the figure is 18–19 per cent). In no developing region do women hold more than 8 per cent of ministerial positions. This shows how the *voices* of the women have been suppressed.

Gender inequalities impose large costs on the health and the well being of men women and children and affect their ability to improve their life. In addition to these personal costs, gender inequalities reduce productivity in farms and enterprises lowering prospects for reducing poverty and ensuring economic progress. Gender inequality also weakens a country's governance and thus the effectiveness of the development policies. The toll on human lives is a toll on development since improving the quality of the people's lives is development's ultimate goal. Gender inequalities also impose costs on productivity, efficiency and economic progress by hindering the accumulation of human capital at home and in the labour market. Actually it excludes women or men from access to resources, public services or productive activities. Gender discrimination thus diminishes an economy's capacity to grow and to raise the standards of living.

The World Bank has proposed certain measures in this regard to remove gender inequality in fostering development. It actually involves a *three way approach*, namely: (i) reforming institutions to establish equal rights and opportunities for women and men; (ii) fostering economic development to strengthen incentives for more equal resources and participation; (iii) taking active policy measures to redress persistent gender disparities in command over resources and political voice. In addressing the above issues some important aspects which need to be taken care of are: (i) *ensuring equality in basic rights*–legal, social and economic rights provide and enabling environment in which women and men can participate actively in the society to attain a basic quality of life and take advantages of the new opportunities that development affords. Greater equality rights is also consistently and systematically associated with greater gender equality in education, health and political participation,; (ii) *establishing incentives that discourage discrimination by gender*—the structure of economic institutions also affects gender equality in important ways. Markets embody a powerful set of incentives that influence decisions and actions for work, saving, investment and consumption. The relative wages of men and women, the returns to productive assets and the prices of goods and services are all largely determined by the structure of the markets; (iii) *designing service delivery to facilitate equal access*—the design of the programme delivery such as education, health, access to finances can facilitate or inhibit equitable access for females and males. Inclusion of the society as a whole becomes important because the need of the men and women can then be realized. The programme would then be endeavoured towards the needs of the people; (iv) *promoting gender equality in access to produce resources and earnings capacity*—fair and equal distribution of productive resources and employment opportunities can advance gender equality and enhance economic efficiency; (v) *reducing the personal costs to women of their household roles*—in the developing countries the women both in the childhood and the adulthood are constrained by the responsibility that they share in the household activities. These have detrimental effects to the women's participation in the education and the market work; (vi) *strengthening women's political participation and voice*; and (vii) *promoting gender appropriate social protection*.

In line with these propositions set by the World Bank, the United Nations Economic and Social Council had brought forward the concept of *Gender Mainstreaming*. The concept of bringing gender issues into the mainstream of the society was clearly established as a global strategy for promoting gender equality in the 'Platform for Action' adopted at the United Nations Fourth World Conference on Women, held at Beijing (China), 1995. It highlighted the necessity to ensure that gender equality is a primary goal in all area(s) of social and economic development. *"Mainstreaming a gender perspective is the process of assessing the implications for women and men of any planned action, including legislations, policies or programmes in all areas and at all levels. It is a strategy for making women's as well as men's concerns and experiences an integral dimension of the design, implementation, monitoring and evaluation of policies and programmes in all political, economic and social spheres so that women and men benefit equally and inequity is not perpetuated. The ultimate goal is to achieve Gender Equality".* Mainstreaming includes gender-specific activities and affirmative actions, whenever women or men are in a particularly disadvantageous position. Gender-specify interventions can target women exclusively, men and women together, or only men, to enable them to participate and benefit equally from the development efforts. These are necessary temporary measures designed to combat the direct and indirect consequences of the past discriminations.

The Basic Principles of Mainstreaming

According to Carolyn Hannan, Director of the UN Division for the Advancement of Women the responsibility for implementing the mainstreaming strategy is system-wide and rest at the highest level within the agencies. Other principles include:

- Adequate accountability mechanisms for monitoring progress need to be established;
- The initial identification of issues and problems across all area(s) of activity should be such that gender differences and disparities can be diagnosed;
- Assumptions that issues and problems are neutral from a gender equality perspective should never be made;
- Gender analysis should always be carried out;
- Clear political will and allocation of adequate resources for mainstreaming, including additional financial and human resources if necessary, are important for translation of the concept into practice;
- Gender mainstreaming requires that efforts be made to broaden women's equitable participation at all levels of decision making;
- Mainstreaming does not replace the need for targeted, women specific policies and programmes or positive legislation, nor does it do away with the need for gender units or focal points.

Inspite of having set all such policy framework in our mind the above propositions should be critically reviewed before being implemented. In offering a critique of the Bank's gender strategy, it is necessary to examine some of the premises on which it is based. Thus the relevant question may be how does the process of economic growth and the growth strategies supported by the Bank affect gender outcomes in terms of equality and welfare? In particular, how do the Bank's fundamentals in terms of macro-economic and sectoral policies affect gender outcomes? These fundamentals as we know calls for privatisation of the State Owned Enterprises (SOEs), trade liberalisation, deregulation of the economic activities and the greater reliance on the market forces to promote greater economic efficiency and higher economic growth.

In many countries however, privatisation policies and closure of SOEs have led to widespread unemployment and reduction in income. This has resulted in greater domestic violence and lesser opportunities for the children of these families in terms of education and health facilities. This is true for most of the developing countries and for also those countries which are undergoing a transition from socialistic to capitalistic structures. Given this context, it is questionable whether sectoral goals in education and health with efforts of reducing gender disparities are really feasible. Gender violence on the other hand cannot be reduced simply through laws or social interventions. These consequences of the Bank's policies in fact drive home very strongly that gender is a cross-cutting issue and cannot be addressed in isolation. Socialistic countries have in addition faced a massive withdrawal of the State and curtailments of the benefits for the women which have hitherto ensured their participation in the labour force on equal terms with men. Globalisation has on the other hand led to strides in terms of female employment, mostly in the light manufacturing industries in the export promoting zones (EPZs). Greater international competition has however meant that much of the labour in the industry is casualised, benefits reduced and working conditions poor. Further changes in global trading rules and increased competitiveness threaten the gains of female employment itself.

Secondly, one may also question as to how the Bank's fundamentals relate to income distribution and thus to gender outcomes, given that it is related to gender inequality. It is evidenced that private sector led initiatives has generated a pattern of growth and development with a more unequal distribution of income in developing countries in the 1990s. Since there are no policy measures to promote wider social equality, one would conclude that in terms of the Bank's own findings this jeopardizes advances in terms of gender equality.

The third area of discrepancy which might emerge in the Bank's policy recommendation regarding gender is that it is country specific, i.e. it should take into account the cultural and the social contexts of the different countries. This has an important face value but is in direct contradiction with the common macro-economic policies—the fundamentals of privatisation followed by the Bank across all developing countries. Country specific approach should start first by considering the consequences of measures such as structural adjustment on poverty, growth and income distribution in borrowing countries. There should be conformity between the two policy approaches such that macro and financial policies are not engendered. The gender outcomes are thus to be identified before implementing macro policies which should also be country specific.

Fourth, it would be an individualistic approach if the rights, resources and voice framework is considered and this is compatible with the Neo-liberal philosophy propounded by the Bank. With the State, as in several developing countries moving away from resource ownership, reducing private expenditure or withdrawing subsidies income distribution and welfare is getting worsened. There have also been major changes in terms of rights are to be promoted and those that are to be curtailed. There has been a significant erosion of the rights of the trade unions and the rights at the workplace of men and women. Privatization leaves a very narrow space for these rights to exist. An evaluation of the impact of changes in the role of the state is thus essential for analyzing gender disparities and gender advances. Given the context and absence of such an analysis, certain measures are likely to succeed and others to fail. For instance, equal rights of men and women to training, schooling etc., will reduce gender segregation of the labour market and also gender differentials in terms of wages. A much wider impact in terms of working conditions may however require a much greater adherence to labour rights and a strong labour movement. Hence the individual rights are not the sine qua non for ensuring social justice and justice between men and women. It is here that Bank's macro policies and the industrial policies, with emphasis on the labour market flexibility may come in conflict.

The final point one would like to make is on the issue of partnership. While it is necessary for the Bank to build partnerships with the developmental agencies working for provision of empowerment, it is an irony that in the present day context, the overall partnership of the Bank with the developing countries is filled with conditionalities which restricts the periphery of the policy making and implementing roles of the States.

However there are ways and means whereby the *Bank's policy measures can be taken a step ahead to foster gender equality*. This includes the use of the following:

Macro-economic indicators should not only be including the variables of the growth rates of GNP and the price level but also should cover the welfare aspects and measurements. Indicators should not be used for the stabilization objectives but also should include social and other objectives. Thus while the goal is to keep prices low and the GNP high, one needs more detailed analysis of the structure of prices and how they affect the poor men and women. The pattern of the GNP growth and the extent to which it is pro-poor offer opportunities to women and the disadvantaged group to come under the area of concern. A Nation should have the scope of deciding on the extent of private –public ownership, the content of trade, agricultural and industrial policies and on the content of the social expenditures.

There should be a greater focus on *mechanisms* and *processes* which would reduce the gender disparities so that appropriate policy interventions can be made. A similar effort can also be made to understand *how expenditure in different sectors is related to outcomes*. The needs of women can then be prioritized and budgets can be prepared.

The conclusion from the above study is that the Bank has to address wider issues of social inequality and justice before it tries to incorporate the issues of gender equality within the framework of development. The pattern of growth propounded by the Bank often leaves aside the issues of social justice. The Bank's macro policies will only become effective when social motivations are embodied with it. We should focus ourselves to the mechanisms and outcomes of policies while addressing the gender issues.

REFERENCES

Arya, Anita, (1963): *Indian Women*; Gyan Publishing House, New Delhi.

Drèze. Jean and Sen Amartya (2002): *India: Development and Participation*; Oxford University Press, New York.

Mazumdar. Vina (ed), 1978: *Role of Rural Women in Development*. Allied Pub. Pvt. Ltd. Mumbai.

Miller, B.D. (1981), *The Endangered Sex*. Ithaca: Cornell University Press, Cornell.

Murthi, et al. (1991), "Mortality, Fertility and Gender Bias in India: A District Level Analysis", *Population and Development Review*, 21 (4): pp. 745-782.

National Policy for the Empowerment of Women 2001, Department of Women and Child Development, Ministry of HRD, Govt. of India, p. 4.

Saha, Shelley and Sundari Ravindran (2002) Gender Gaps in Research on Health Services in India. Journal of Health Management, Vol. 4, pp. 185-214.

Sahay, Aparna (2004) Gender Disparity and Women's Autonomy. Prashasanika, Vol. XXXI, No. 1, pp. 85-92.

Soundari, M. Hilaria and M.A. Sudhir (2003) Gender Disparity in Tamil Nadu. Social Welfare, Vol. 49, No. 12, pp. 24-27.

Sudharani, K. et al. (2003) Awareness of Women About Their Rights. Social Welfare, Vol. 50, No. 4, pp. 20-22.

3

Gender and Development
The Ways Ahead?

Miss. Sancheeta Ghsoh

"The real wealth of a nation is its people. And the purpose of development is to create an enabling environment for people to enjoy long, healthy and creative lives".

United Nation Development Programme, 1990.

The milieu of integrating women in the development process is now engaging the attention of planners and policy-makers all over the world, not so much because of the International Women's Decade and the United Nation's call for action plans for these purposes but because of increasing realization of social imbalances that development itself has created.

Women constitute at least half of the world's population and undertake the major number of work hours. However, the return she gets in comparison to men is very meager. The phenomenon of discrimination against women cuts across geographical boundaries or the levels of development. Not only in developing countries, but the discrimination against women exists in many other parts of the world though less severe. Over the last half century, there has been a significant effort to promote equal rights and opportunities for both women and men. But the magnitude of the problems still waits to be explored threadbare. One of the most important determinants of gender inequality in the developing countries mainly in India is the patriarchal nature of society, which leads to widening the gap of gender inequality.

Gender, which refers to the social classification of men and women into masculine and famine emphasizes mainly on social roles and responsibilities of men and women. But from the very beginning it resulted into domination of men over women in every sphere, economy, production, education and so on. In addition to basic inequalities, in access to education and resources and an unequal share of the burdens of the poverty, women continued to be under-represented in formal decision-making structures.

Why has Gender Become So Important in Development Issues?

Recognition of the need to improve the status of women and to promote their potential roles in development is no longer seen as an issue of human rights or social justice. While the pursuit of gender equity remains strongly embedded within the framework of fundamental human rights and gender justice, investments in women now also are recognized as crucial to achieving sustainable development. Study shows that, low level of education, poor health and nutritional status, and overall limited access to resources not only depress women's quality of life, but also limit productivity and hinder economic efficiency and growth of the society as a whole.

Women's or gender concerns were brought to focus when it started becoming clear that planned development efforts, which were meant to improve the lives of the whole community, were either not helping women, or were actually harming them in some way or other. Around the end of early 1960s and early 1970s, women researchers in different parts of the world started pointing to the neglect of women in development planning. Planners assumed that development programmes would automatically help all the members of the community, but this assumption was found to be invalid in almost everywhere. In the 1950s, when newly formed independent countries began planned development, the model was adopted from the west. It was thought that industrialization and modern agriculture would usher in growth and development, and therefore, the focus was on industrialists, landowners, rich farmers and entrepreneurs. Little attention was paid to the poor and to the women. Women's contribution to the household income and to the economy was neither recognized nor valued. It was discovered that even when a household benefits from any development programme, it does not follow that the women in the household will benefit equally or benefit at all. Studies done from a feminist perspective in different parts of the world provided data and evidences to show the *gender blind development plans* had generally ignored women, their needs and interests. This neglect points out that existing inequalities between men and women were not addressed and second, women's action and potential contribution to and participation in the development process were ignored.

The early 1970's model of 'integration', based on the belief that women could be brought into existing mode of benevolent development without a major restructuring of the process of development, has been the objective of most feminist critique. The alternative vision put forward of 'development with women', demands not just a bigger piece of someone else's pie, but a whole new dish (Momsen, 1993). International development has been challenged to transform itself into a process that is both human-centered and environmentally conservationist.

Along with the traditional challenges of achieving greater improvements in women's health, education, access to financial services, and income-generating opportunities, there are new challenges emerging in the region that need to be addressed in case of gender concern in development. The challenges of dealing with the increasing feminization of poverty, increasing spread of human immunodeficiency virus/acquired immunodeficiency syndrome (HIV/AIDS) among women, the trafficking in women and girls, problems of migrant women workers, violence against women, and female infanticide were brought to the forefront by the women of the region at both the Jakarta Regional Meeting (1994) and the Beijing World Conference on Women (1995). Many of these new challenges and emerging areas of concern are conspicuously manifested in the region. Feminization of poverty is a serious concern in Asia, where two thirds of the world's poor reside, of which two thirds are women. In poor households, women are more disadvantaged than men in terms of their workloads and access to resources and remunerative activity. The phenomenon of migrant women workers, who are recognized as a vulnerable group in the Beijing Platform for Action, is increasing in the Asian and Pacific region. Migrant women workers contribute substantially to the national economies of several countries and are the mainstay of the remittance economy relied upon by many poor urban and rural families. In spite of the valued contribution of migrant women workers not only to their families but also to the economy, there is an absence of protective legislation and regulatory mechanisms to prevent their exploitation and ensure their protection. Households headed by women are particularly vulnerable. Disproportionate number of women among the poor pose serious constraints to human development because children raised in poor households are more likely to repeat cycles of poverty and disadvantage. Disproportionate burdens of structural adjustment and economic transition are borne by women. For example, in the Central Asian countries, the social and economic advancement of women fostered by public policies and expenditure in the past is being slowed or reversed as a result of restructuring. Likewise, some of the austerity measures being pursued by many of the developing nations under adjustment programmes impact more negatively on women than men. Asia has the second highest prevalence of HIV infection among women. On a world scale, South Asia is the highest after sub-Saharan Africa in the degree of prevalence of HIV infection among women. In Asia, almost half of all adults newly infected with the virus are women compared with less than 25 per cent just 6 years ago. In the year to mid-1994, Asia's share of global AIDS cases rose eightfold. Using Cambodia as an example, a 1994 World Health Organization (WHO) study indicated that nearly 40 per cent of sex workers surveyed in Sihanoukville were infected with HIV. There were concerns that the virus had already gone beyond this risk group into the general population. The nexus between the spread of HIV, and the low status and powerlessness of women, is seen as one of the major contributors to the spread of HIV/AIDS among women in the region. Gender-selective abortion, female infanticide, trafficking in women, abuse, and malnutrition among girls seriously compromises the rights of the girl child in some. Violence (homicides, domestic violence, rape) against women, although widespread in all cultures, ages, and income groups, has ill effects particularly on the well-being and productivity of women mainly in the developing world, more specifically in India.

Concept of Engendered Development

It was thought that if the developmental issues focus on gender inequality it can help women from deprivation. That is why the concept of *engendered development* has come into lime light. The Constitution of India not only provides for equal rights and privileges for women and men but also for making special provision for women. A series, of social legislations have been enacted from time to time for raising the status of women in the country. The Five Year Plans have consistently placed special emphasis on providing minimum health facilities integrated with family welfare and nutrition for women and children, acceleration of women's education, their increase in the labour force and welfare services for women in need. Gender sensitive Census (2001), gender budgeting, various welfare and development schemes have been introduced to improve the living conditions of women and to increase their access to and control over material and social resources. Special steps have been taken to remove legal, social and other constraints to enable them to make use of the rights and new opportunities becoming available to them.

Gender Mainstreaming: 'Gender mainstreaming' was defined by the United Nations Economic and Social Council in 1997 as 'a strategy for making women's as well as men's concerns and experiences an integral dimension of...the policies and programmes in all political, economic and societal spheres so that women and men benefit equally and inequality is not perpetuated.'

Indian Five Year Planning and its Gendered Development Aspect

The real development can not take roots if it bypasses women, who not only represent nearly half of the country's total population but also represent the very kernel around which societal reorientation takes place has been the guiding principle in the formulation in our plans from the very beginning (Arya, 1963).

India's five year planning started approaching women's development as welfare programme from the very beginning of first plan. It includes high priority to education of women and introduced measure to improve maternal and child health services. The Planning Commission of India has been addressing the concerns of women particularly, from the Sixth Five Year Plan onwards when a chapter on Women and Development was added to the document. It has also been holding consultations with women's groups for their inputs. The eighth plan marked a shift in development to empowerment in approach to various women development schemes. The ninth plan approached to two major steps towards gender justice and new approach of women empowerment. However, for engendering the Ninth Five Year Plan, an organized initiative was supported by UNIFEM. It worked in tandem with the Government, in particular with then the Department of Women and Child Development and the Planning Commission and involved some of the most distinguished leaders of the women's movement and grassroots women's group through the creation of a Think Tank. These schemes are continuing in the tenth plan also. It was for the first time that women's voices from different walks of life and different regions were fed into the planning process at different stages.

Concern for Women's Development—Declining Sex Ratio and *Female Missing* in India

Inequality between men and women is one of the crucial disparities in many societies, particularly in India. In many countries, women tend in general to fare quite badly in relative terms as compared with men, even within same families. This is related not only in such matters as education and opportunity to develop talents, but also in more elementary fields of health and nutrition (Amartya Sen, 1992).

One result of this female deprivation is remarkably low ratio of females to males in the Indian population as compared with the corresponding ratio in Europe or North America. Even though male outnumbers female birth, in the developed countries like Europe and North America, females tend to outnumber males and the ratio is around 1.05 (United Nation Population Division, 1999).

In contrast many parts of the developing countries have female male ratio below unity, e.g. 0.98 in North Africa, 0.95 in Bangladesh, 0.94 in China. And in India the female male ratio is much lower than these countries, which is 0.93 (Dreze and Sen, 2002).

Therefore, much evidences come from India, which reflect a sharp deficit of women, the situation termed by Amartya Sen as *missing female.* This results in relative neglect of the health and well-being of women (particularly young girls and female infants) resulting in a survival disadvantage of females than males.

Development Programmes Leading to Marginalization of Women

For development to be meaningful to women it must first recognize the existing inequalities in society, inequalities which have been brought about by class oppression, and inequities specific to women that result from gender oppression. Women carry the burden of poverty. And yet historically, development for all its grandiose aims has marginalized women. She has been neglected and further marginalizaed in the field of agriculture, rural development, access to employment and so on. The experience of different parts of the world shows that women have been pushed out of mainstream agriculture in the name of 'development'. Earlier, men and women were equal partners in agriculture. Their knowledge, contribution and participation in decission making were more or less the same. Gradually in the course of time, the male farmers singled out for attention by male 'developers'. Most training, information and credit for agriculture, horticulture and animal husbandry have been given to male farmers, in spite of the major contribution of women to these activities.

Not only in agriculture but also in other spheres of developmental activities, women remain in the last row due to blind and faulty planning and policies. For example, in India it has been found that as a result of mechanization and modernization women lost their jobs in the textile industry where they had been employed in large numbers. This economic marginalization has led to women's social marginalization to a lowering of their status. This may be one reason, why dowry and female infanticide and foeticide in India have spread to areas and communities in which they did not exist earlier. This may also be one of the main reasons for continuing decline in the female-male ratio in the South Asian countries, particularly in India.

Women and Environment: Though the environment is gender free, the social bariers created in the society make women greater victimes of environmental problems than men. The environmetal degradation due to modernization and development affected women more worstly than men. The deterioration of natural resources displaces communities, especially women, from income generiting activities while greatly adding to unenumerated work. In both urban and rural areas, environmental degradation results in negative effect on health, well being, quality of life of the population at large, especially girls and women of all ages.

In case of political representation in India both in national parliament and state assemblies women again are marginalized. In spite of having 33 per cent reservation in political representation and in *Panchayati Raj* Institution, according to the 73rd constitutional amendment bill, shadow area still persists. Power are not given in true sense to the women in need and again men are in the steering position.

Another problem arises due to globalization and development is marginalization of women in urban formal sector employment. Female are also facing problem if they are migrant to any city or new work place. Increased participation of women in the labour force, while providing women with much-needed access to income, has in some instances generated new problems for working women. Issues such as poor working conditions, exposure to health risks, higher incidence of industrial diseases, worker health and safety, and new forms and patterns of exploitation such as sexual harassment in the workplace, wage differentials with men in same position are new areas of concern.

The argument of women's deprivation can be continued with the situation in health care. Continuously women have been neglected in access to primary health care. Discrimination against female child for health care is a very common phenomenon in both rural as well as in urban areas. India is among the few countries where female mortality from infancy to nine years of age exceeds that of males. Maternal mortality in India is among the highest in the world, at more than 600 maternal deaths per thousand live births (Ravindran, 1993).

Conclusion

But only by undertaking these concerted strategies can the government, donors and civil society begin to address the issue of female deprivation in development? Women are among the worst affected by the crisis-ridden development strategies adopted by the country. After four decades of development efforts, gender gaps in access to education, employment and health care provision have persisted and are closing extremely slowly. Even today, more than two thirds of the female population is illiterate, as compared to only one third of the male population. The recent population policy also failed to capture the real problems of women. The new and emerging areas of concern for women described and internationally endorsed in the Beijing Platform for Action are acknowledged as important development issues. Beside questions of equity and rights, these problems are recognized as infringing on women's ability to participate and contribute to development as a whole. Attention is focused on the social, political and economic costs of all violence against women. The relationship between female-focused violence and maternal mortality, health care utilization, child survival, AIDS prevention, costs to the judiciary and law enforcement agencies, other economic costs and socio-economic development are receiving increasing attention. Above all it is not only an issue of fundamental justice, equality and human rights, but also an important public health and development issue for communities and governments.

Trying to develop the society without acknowledging the people who do two thirds of the work will obviously invite failure. Ways must be found to reduce their burden of work if women are to realize their potential. Women are the agents of development not the victims. Let women be empowered to take the major portion of the process of development which is for the women.

REFERENCES

Arya, Anita, (1963): *Indian Women*; Gyan Publishing House, New Delhi.

Drèze. Jean and Sen Amartya (2002): *India: Development and Participation*; Oxford University Press, New York.

Mazumdar. Vina (ed), 1978: *Role of Rural Women in Development*. Allied Pub. Pvt. Ltd. Mumbai.

Momsen. J. H, (1993): *Women and Development in the Third World*; Routledge Publication, London.

Ravindran. T.K.S, (1993): "Women and the Politics of Population and Development in India", *Reproductive Health Matters*, Vol. 1.

Sen, Amartya (1992): "Missing Women", *British Medical Journal*, Vol. 304.

4

Gender Issues in Development
A Critical Approach

Miss. Sailaja Nandini

Gender issues are not the same as women's issues. Understanding gender means understanding the opportunities, constraints and the impacts of change as they affect both men and women".

The World Bank

Introduction

Gender stratification is evident in every culture and society in the world. Some things are masculine others are feminine, some work is women's work, some responsibilities are women's responsibilities, even major religions of the world assign different social responsibilities to men and women. The differential evaluation of people's social worth primarily on the basis of sex is the key aspect of gender stratification. Such views become a characteristic of the entire social system leading to unequal distribution of power, prestige and property. Gender inequality affects every aspect of culture and society. Its effect is most prominent in family structure, the education system and the economy. Just like social class systems, gender is a structural feature of society.

As a result of gender stratification the universal status of men is higher than women. Therefore, men enjoy a greater allocation of societal resources of power, prestige, and property. Since the status of women is universally subordinate to men in the society, women in the United States and India are no different. However, variations on the basis of the institutional spheres are obvious when we compare the status of women in the United States and women in India. In the United States and India different structures relating to family, politics, religion, law and economic system exist. Moreover, both countries have different cultural heritages and religious backgrounds. India has a distinct culture and traditions, while America is striving towards cultural integration, modernization and liberalization. In recent years India has made efforts to become a modern and liberal state, free from its traditional past, but such efforts have been undermined by the exponential growth in population and poverty.

Gender and development is becoming an accepted perspective in practical development work. A focus on gender issues means looking at both men and women, whilst recognizing that it is women who suffer from gender inequality and discrimination. Consideration of gender issues is now common currency in government, NGOs, the private sector and mainstream politics in India, at least at the rhetorical level. This awareness has come about as a result of the efforts of the women's movement and the influence of international development and feminist debates, as well as aid initiatives with a focus on gender. However, this increased awareness does not imply that gender issues are being dealt with on their own terms.

Recognition of the need to improve the status of women and to promote their potential roles in development is no longer seen only as an issue of human rights or social justice. While the pursuit of gender equity remains strongly embedded within the framework of fundamental human rights and gender justice, investments in women now also are recognized as crucial to achieving sustainable development. Economic analyses recognize that low levels of education and training, poor health and nutritional status, and limited access to resources not only depress women's quality of life, but also limit productivity and hinder economic efficiency and growth. Hence, promoting and improving the status of women need to be pursued, for reasons of equity and social justice and also because it makes economic sense and is good development practice. The range of actors with a stake in promoting 'women's' issues each have their own underlying agenda which influences the way in which gender questions are addressed. Moreover, analysis of gender issues cannot be separated from broader political and economic developments. Gender relations are being continuously reshaped by contemporary developments so that it is necessary to look at the complex ways in which gender relations are being recast today. In many societies, women are denied access to basic services and essential assets such as land, and excluded from decision-making. One-thirds of the world's poor live in this region, the majority of whom are women.

Over the last decade, Gender and Development has gained a space in development projects and programmes—at least at the level of policy formulation. Talking of Women and Development (WID) is outdated—to day the catchphrase is Gender and Development: GAD. This change has been pushed by women's movements on a global scale, and it has gained momentum especially since the Fourth Wold Conference of Women in Beijing 1995. Even in Mozambique a Post-Beijing in Plan of Action was issued in 1996, specifying GAD policies ministry by ministry.

Women and Development

- 1950s to 1960s—Women's issues were seen mainly within the context of human rights.
- 1970s—the formula was to "integrate women into development."
- 1980s—The United Nations' Third Development Decade gave rise to a "trend towards seeing women as equals, as agents and beneficiaries in all sectors.

SOME GENDER ISSUES IN DEVELOMENT

Poverty

Women now constitute 70 per cent of the world's 1.2 billion poor. Women are especially vulnerable to poverty. Women and girl children suffer from gender discrimination in the allocation of resources within the household, in spite of their considerable labour and often cash contributions. This discrimination is particularly marked in the allocation of food and health care resources, resulting in imbalances in the sex ratio for most states. In general, where women are productive work is not visible, or where gender differentials in earnings are high, women may be particularly prone to discrimination in the household.

Women's access to land, particularly agricultural lands often the determinant of their income status. Poor women have no access to land resources. Women largely constitute the unskilled labour force in both agricultural and non-agricultural labour markets. So far as the land reforms has not given any cognizance to the existing gender inequalities in land inheritance laws and ceiling laws.

Education

Over two-thirds of the world's 1 billion illiterate are women. An overview of the literature that examines education in India reveals that there is a huge gender disparity. While in the last decade there has been a significant increase in the country's literacy rates–a 13.7 per cent increase from 1991-2001 (IB), especially in the literacy rates of women, they are still the disadvantaged sex and are at the lower end of the spectrum- 51.4 per cent literate women: 74.5 per cent literate men. Women in India, on the average, are paid less than 40 per cent of what men are paid. Sex ratio in India is among the poorest in the world; there are over 100 million fewer women in India. Female infanticide continues to be amongst the highest in the world. Both economically and health wise, women in India, thus, find themselves systemically or socially discriminated against more than anywhere else in the world. Studies claim that this is closely tied to literacy; figure 5 shows that infant mortality rates are closely tied with the literacy levels in women. In India, however, women were among the lowest literacy rates. Enrollment level of women in schools is more than 20 per cent lower; adult literacy levels are more than 30 per cent lower.

Health

Gender influences both health status and health care. Women are the fastest growing group of HIV infected adults. Each year at least half a million women die from complications due to pregnancy and another 700,000 due to unsafe abortions. Sex ratio is less unequal among the poorest classes. Also, women may be more susceptible than men to diseases, which cause death. Comparisons of the morbidity of men and women in the same households usually show female morbidity to be higher, possibly due to lack of health care. One of the important causes of the deteriorating sex ratio is female feticide or sex selective abortion. Women's health is also affected by their work.

Both gender differences and gender inequalities can give rise to inequities between men and women in health status and access to health care. For example: A woman cannot receive needed health services because norms in her community prevent her from travelling alone to a clinic. A married woman contracts HIV because societal standards encourage her husband's promiscuity while simultaneously preventing her from insisting on condom use. Malnutrition in India has exceptionally high rates of child malnutrition, because tradition in India requires that women eat last and least throughout their lives, even when pregnant and lactating. Malnourished women give birth to malnourished children, perpetuating the cycle. Poor Health Females receive less health care than males. Many women die in childbirth of easily preventable complications. Working conditions and environmental pollution further impairs women's health.

Violence

Violence against women is an obstacle to the achievement of the objectives of equality, development and peace. Violence against women both violates and impairs the enjoyment by women of their human rights and fundamental freedoms. The term violence against women means any act of gender based violence that results in physical, sexual, or psychological harm or suffering to women including threats of such acts, coercion or arbitrary deprivation of liberty, heather occurring in public or private life. Acts of violence against women also include forced sterilization and forced abortion, coercive/forced use of contraceptives, female infanticide and parental sex selection. Violence against women is a manifestation of the historically unequal power relations between men and women, which have led to domination against women by men.

Violence against women forms the core of gender-based inequalities with far reaching consequences for women's development. Gender violence encompasses violence against women within the family or within the general community—including rape, sexual abuse, sexual harassments and forced prostitution, and violence perpetrated or condoned by the state. In the US, a woman is physically abused every 8 seconds and one raped every six minutes. In India one woman dies of dowry in every 102 minutes, rape of one woman occurs in every 54 minutes, in every 51 minutes sexual harassment of one woman takes place, a woman is being kidnapped in every 43 minutes, in every 33 minutes a woman is victim of cruelty, in every 26 minutes, a woman is molested; in every 7 minutes she is a victim of criminal offence (including domestic violence).

Armed and Other Conflicts

Women often have no decision-making power during global conflicts. Women constitute 80 per cent of the world's 23 million refugees and other displaced persons. While entire communities suffer the consequences of armed conflict and terrorism, women and girls are particularly affected because of their status in society and their sex. Parties to conflict often rape women with impunity, sometimes using systematic rape as a tactic of war and terrorism. The impact of violence against women and violence of the human rights of women in such situations is experienced by women of all ages, who suffer displacement, loss of home and property, loss or involuntary disappearance of close relatives, poverty and family separation and disintegration, and who are victims of terrorism, murder, torture, sexual slavery, rape, sexual abuse, and forced pregnancy in situations of armed conflict.

Economic Participation

In many regions, women's participation in remunerated work in the formal and non-formal labour market has increased significantly and has changed during the past decade. While women continue to work in agriculture and fishries, they have also been increasingly involved in micro, small and medium–sized enterprises and, in some cases, have become more dominant in the expanding informal sector. Due to inter alia, difficult economic situations and a lack of bargaining power resulting from gender inequality, many women have been forced to accept low pay and poor working conditions. Women migrant workers, including domestic workers, contribute to economy of the sending country through their remittances and also to the economy of the receiving country through their participation in the labour force. However, in many receiving countries, migrant women experience higher levels of unemployment as compared with both non-migrant workers and male migrant workers. Lack of employment in the private sector and reductions in public services and public service jobs have affected women disproportionately. In some countries, women take on more unpaid work, such as the care of children and those who are ill, compensating for low household income, particularly when public services are not available. Although some new employment opportunities have been created for women as a result of the globalization of the economy, there are also trends that have exacerbated inequalities between men and women.

Power Sharing and Decision-Making

More than 100 countries have no women in the government. Negative stereotypes contribute to the discrimination that women face. Despite the wide spread movement towards democratization in most countries, women are largely underrepresented at most levels of government, especially in ministerial and other executive bodies. Women in politics and decision making positions in Governments and legislative bodies contribute to redefining political priorities placing new items on the political agenda that reflect and address women's gender specific concerns, values and experiences, and providing new perspectives on mainstream political issues. The underrepresentation of women in decision-making positions in the areas of art, culture, sports, the media, education, religion and the law have prevented women from having a significant impact on many key institutions.

In case of local level participation such as in countries like India women's participation in Panchayati Raj is a key issue of concern. Though women are coming to the decision-making level through 33 per cent reservation in Panchayati Raj but still they don't have really the power. Their male counterparts actually taking the decessions.they are in the decision making level but only at pen and paper.

Mass Media

More women are involved in careers in the communications sector, but few have attained positions at the decision-making level or serve on governing boards and bodies that influence media policy. The lack of gender sensitivity in the media is evidenced by the failure to eliminate the gender based stereotyping that can be found in public and private local, national and international media organizations. Also the major issue in development of women is the portrayal of stereotype roles and indecent representation of women in the electronic and print media.

Environment and Gender Issues

The link between gender issues, the environment and development policy has intensified during the last decade. Women have traditionally been responsible for subsistence and survival tasks like providing water and food, fuel and fodder collection. Thus their dependency on the environment is very crucial; being deeply linked with their survival. Degradation of Environment directly affects women's deprivation.

The Girl Child

In many countries available indicators show that the girl child is discriminated against from the earliest stages of life, through her childhood and into adulthood. They are victims of so many violence such as early marriage including child marriage, sexual abuse, discrimination against girls in food allocation, female infanticide, female foeticide, girl child labour and other practices related to health and well being.

This is linked to the shortcomings in health systems, poor nutrition, the cost of health care, the lack of infrastructure and skilled personnel, and persistent cultural attitudes and practices–such as son preference and limited contraception–that continue to impose a cost on women's reproductive health. The epidemic proportion of gender-based violence underscores women's overall vulnerability, the persistence of cultural norms, and the bias of current legal systems, the despair associated with poverty, skewed power relations, and the limited protection that exists for women. The increase in the traficking of women and girls as a result of more integrated markets and the increase in labour migration opportunities. Representation of women in decision-making in East Asia falls short of meeting MDG targets. Barriers continue to reinforce unequal gender relationships and perpetuate women's marginalization in the decision-making sphere. Part of this challenge is political, requiring policies and frameworks that would enable women's increased participation. The other part is cultural; demanding a change in mind-set that would challenge the current attitudes and practices that obliterate women's potential as a decision-making partner. Getting laws and institutions in place is undoubtedly an important step for women in East Asia. However, Impact is limited unless these can be implemented and used effectively. Until then, traditional rights of marriage and inheritance will continue to undermine women's access to land in many parts of the region, violence against women will continue to increase, and women will remain vulnerable in the workplace. And in a region of large-scale labour mobility, the opportunities for human trafficking will increase almost unchecked.

Conclusion

Gender issues in development are a major area of concern. In addressing inequalities in health status and unequal access to and inadequate health care services and educational services, unequal power and decision making level between men and women, indecent representation of women in media, and to face the feminization of poverty, enhance the work participation of women, governments and other actors should promote an active and visible policy of mainstreaming a gender perspective in all policies and programmes.

REFERENCES

Fourth World Conference on Women (1995) "Platform for Action", Beijing, China.

Engendering Development: Through Gender Equality in Rights, Resources and Voice. Washington, DC: World Bank, 2001 Bennett, L. 'Women, Poverty and Productivity in India'. World Bank EDI Seminar Paper 43. Washington, DC: World Bank, 1992.

Elson, D. Male Bias in the Development Process. Manchester: Manchester University Press, 1991.

Marcoux, A. 'The Feminization of Poverty: Claims, Facts, and Data Needs'. Population and Development Review 24(1) (1998): 131-9.

Murthi, M., Guio, A. and Dreze, J. 'Mortality, Fertility and Gender Bias in India: A District Level Analyses. Population and Development Review 21(4) (1995): 745-782.

5

Methodological Issues in Gender Budgeting Initiative (GBI) in India

Mr. Bibhu Prasad Sahu

"To Awaken the People, it is the women who should be awakened. Once she is on the move, the family moves and the nation moves."

—Pandit Jawaharlal Nehru

Introduction

A budget is not merely a statement of income and expenditure rather it is an instrument for fulfilling the obligations of the state towards its citizens. It is the priorities set by the government in allocating resources in a planned manner to achieve certain desired objective in a financial year. The choice of the period varies from country to country. In India the government transactions normally carried out from 1st April to 31st March. The budget document of the government states the rules and principles on the basis of which the public economy is managed effectively (Musgrave, 1961), The tax and expenditure policies are chosen and given practical shape in the budget.

The size of the budget is alternatively the quantum of the public goods to be produced in an economy and is normally decided by a political process. In such a situation three groups of persons i.e. government officials, the elected representatives and the citizens interact to finalize the basket of public goods to be produced. To a large extent, all these groups are motivated by the desire to satisfy their own preference schedules by influencing the government's decision making. The higher income groups the government officials and the elected representatives represent one and the same and influence disproportionately to their strength even in a democracy to steal a large share of the benefit of public investment (Due and Friedlaender,1973).

This means the budgetary policy is ultimately decided by a bargaining process and the share of different groups is ultimately decided by their bargaining strength. In order to pressurize the government, different pressure groups stage protests, strikes and lockouts, rallies and if necessary self immolation bids to get their share jacked up. Since the details of the budget are prepared mostly by the bureaucracy, the attitude of government officials ultimately is reflected in the pattern of resource allocation. The elected representatives marginally after the allocation after debating in the parliament /legislature.

Over the years, given the situation outlined above, the gender responsiveness of the budget has been under intense debate and the paper is making an attempt to outline the methodological issues involved in it. The paper is divided into three broad sections. Section-I gives the macro-economic analysis of gender budget. In section-II, the experience in international and national level is discussed. Section-III gives the methodological issue involved in the computation of the gender budget.

SECTION-I

Gender Issues in Economic Analysis

The participation of women in an economy has been recognized since long even though there exists sharp differences among the experts on the question of including household work carried out by the female member of a family in the national income. In our country, women constitute slightly less than 50 per cent of the total population and their participation in the workforce is around 26 per cent of the total working population. Assuming 50 per cent of the total population belonging to male and female groups as capable and working, it can be concluded that 24 per cent of the population i.e. 12.5 crore is confined to household works and not exposed to outside work participation

(Census, 2001). Female labour force participation, wage differential, and gender discrimination were issues incorporated into the economic analysis by A.C Pigou, J.R. hicks and more recently by G. Becker, Gary Becker in the sixties emphasized on household economics and used market concepts in household production and time allocation analysis.

Macro Economic Policies and Gender

Macro economic policies generally aim at achieving sustainable economic growth with redistributive justice. The government uses exchange rate policy, monetary policy and fiscal policy combindly to stabilize and restructure the economy. In spite of this there is growing inequality between man and woman at various levels. So the concerns of gender inequality need to be built into macro economic policy framework. Addressing gender concern has dual gains. It not only promotes equality among citizens but also benefits the society through efficiency gains. Empirical literatures have already proved this. Research on agricultural productivity in Africa has shown that given men and woman the same level of agricultural inputs and education, increased yield has come from women farmers to the extent of more than 20 per cent. Similarly a World Bank Study (1995) reported that increases in women's well being yields productivity gains in future. The chance of a child enrolled in a school and getting education increases in case of a woman of a family being educated. The extra income of a woman has also a positive impact on the household's investment on education and health. Therefore there is a need for framing macro economic policies to address gender concerns. Among various policies, budgetary policies particularly the revenue and expenditure policies, are the visible instruments for integrating gender concerns. So a gender budget is not a separate budget for women, rather it is an analysis of government budget with a gender lens. It examines the budgetary needs with its impact on man and woman separately and finds out how gender commitments are translated into budgetary commitments. It stresses on reprioritization of expenditure in a rational way so as to fulfill the needs of man and woman.

Gender Budgeting: The Concept

" 'Gender-sensitive budgets', 'gender budgets', and 'women's budgets' refer to a variety of processes and tools aimed at facilitating an assessment of the gendered impacts of government budgets. In the evolution of these exercises, the focus has been on auditing government budgets for their impact on women and girls. This has meant that, to date, the term 'women's budget' has gained widest use. Recently, however, these budget exercises have begun using gender as a category of analysis and the terminology 'gender-sensitive budgets' is increasingly being adopted. It is important to recognize that' women's budgets' or 'gender-sensitive budgets' are not separate budgets for women, or for men. They are attempts to break down, or disaggregate, the government's mainstream budget according to its impact on women and men, and different groups of women and men, with cognizance being given to the society's underpinning gender relations." (Sharp, Rhonda: 1999)

The gender budgeting exercise would include: (a) Addressing gap between policy commitment and allocation for women through adequate resource allocation and gender sensitive programme formulation and implementation; (b) Mainstreaming gender concerns in public expenditure and policy; (c) Gender audit of public expenditure, programme implementation and policies–relating to public expenditure, fiscal and monetary matters etc.

SECTION-II

International Experience

A fundamental rethinking on macro-economic policy framework is required to address the gender concerns. Like pro-poor budgeting and green budgeting, gender budgeting has immense value for an economy at this juncture. In this regard, the United Nations Conferences on Women generally referred to as Beijing Conference (1995) and the subsequent Beijing platform for action contributed to the emergence of an international consensus in integrating a gender perspective in all policies and their budgetary dimensions.

Given the importance, many countries have initiated this exercise earlier. Australia was the first country, which started this exercise in 1984. But it could not be done on a sustained basis, as the civil society groups did not own the process. The process began in South Africa during 1997 with a participation of a number of agencies. Here the challenge is to develop gender disaggregated inf nation, which is a huge task. In Spain, the government of the Basque county has shown interest in introducing a gender sensitive budget approach. In 2000 there were two initiatives, which moved the process forward. In Sri Lanka, five ministries were selected in 1998 to examine the gender impact of recurrent spending, as well as the gender distribution of public sector employment. One common finding was "that a proper mechanism is to be developed to collect data disaggregated by gender". The procedures and mechanisms identified in this study is to gather information, as a continuous process that will be useful for future policy formulation. So the experience is that gender budget process has to be done over a period of time. Gender budgets are intensely political processes and require strong political will. Over 40 countries around the world have responded to norms of preparing and presenting gender sensitive budgets.

Gender Budget—Indian Context

After India got her independence, planned development approach was considered as the best strategy to raise the growth of the economy and improve the standard of living of the people. During successive five-year plans steps have been taken for the upliftment of weaker sections of society. But concrete steps to empower women were taken during 9th five-year plan. On Feb 19, 1999, the National Development Council adopted a resolution to empower women as socially disadvantaged groups and as agents of socio-economic change. Development of women as a specific objective was incorporated in the Ninth plan (1997-02). A separate "women's Component Plan" like tribal Sub-plan was formulated to ensure that not less than 30 per cent of funds and benefits flow to women from developmental sectors. Accordingly the Planning Commission directed all Central Ministries and Departments and also to the state governments to identify a Women's Component Plan in their respective areas. Specific schemes and programmes in different departments of developmental sector has to be chalked out for this purpose. The Department of Women and Child

Development, in the Ministry of HRD, Government of India in collaboration with United Nation Development Fund for Women (UNIFEM) has initiated gender budget works in India.

The United Nations Development Fund for Women (UNIFEM), Delhi has organized workshops throughout India to create awareness on the issue at all government levels. In 2001-02 taking initiative in this respect, the budgetary allocation on Pro Women Allocation has increased from Rs. 10596.37 crore in 2001-2002 to Rs. 13036.01 crore in 2002-2003 reflecting a percentage increase of 23 per cent. Table-5.1 given next page represents the case.

The table depicts the public expenditure made by the Government of India to increase the allocation to women–related schemes and programmes which is the most important aspect of developing gender justice in India. It is worth noting that the GoI made a total expenditure of Rs. 11204.45 crore in 2001-02 distributed among the Departments of Rural Development, Education and Family Welfare. Then Department of Rural Development has spent a major part of their expenditure amounting of Rs. 4108.62 crore whereas for Education and Family Welfare departments it is Rs. 3478.88 crore and Rs. 1855.86 crore, respectively, in the same year.

As a result of the awareness campaign initiative of the UNIFEM, for the first time, in 2005-06, the budget speech of the finance minister announced gender sensitive budget and made provision on gender lines in case of ten demands for grants in the central ministries. Again in 2006-07 budget, the Finance Minister enlarged the scope of gender budgeting and included schemes where 100 per cent of the allocation is for the benefit of women as well as schemes where at least 30 per cent of the allocation is targeted towards women. The statement now covers 24 demands for grants in 18 Ministries/Departments and five Union Territories and schemes with an outlay of Rs. 28,737 crore. Several Ministries and Departments have initiated an exercise to prepare a public expenditure profile of their budgets from a gender perspective and 32 Ministries and Departments have set up Gender Budgeting Cells.

Table 5.1: Public Expenditure of Government of India with Pro Women Allocation (Rs. in crore)

S. N.	*Ministry/Department*	B.E. 2001-02	R.E 2001-02	B.E. 2002-03	*Percentage Variation* (BE 2001/RE2001	(BE 2001/RE2002
1.	Agriculture and Co-operation	45	44.53	73.88	-1	64
2.	Health	686.7	663.42	708.69	-3	3
3.	Family Welfare	2025.68	1855.86	2932.06	-8	45
4.	Indian System of Medicine and Homoeopathy	49.67	40.45	60.26	-19	21
5.	Education	3359.25	3478.88	4048.03	4	21
6.	Youth Affairs and Sports	94.5	96.64	C 101.82	2	8
7.	Labour	287.15	238.81	240.02	-17	-16
8.	Non-conventional Energy Sources	74.52	63.39	74.62	-15	0
9.	Science and Technology	247.19	237.68	304.28	-4	23
10.	Small Scale Industries and Agro & Rural Industries	276.68	259.1	282.81	-6	2
11.	Urban Employment and Poverty Alleviation	50.4	13.65	31.5	-73	38
12.	Rural Development	3299.39	4108.62	4052.45	25	23
13.	Social Justice and Employment and Tribal Affairs	100.24	103.42	125.6	3	25
	Total	**10596.37**	**11204.45**	**13036.01**	**6**	**23**

Source: Union Budget 2001-2002 and 2002-2003, Government of India.

SECTION-III

Methodological Issues

The Government of India followed the methodology developed by experts of National Institute of Public Finance and Policy (NIPFP) on aggregations of women benefiting expenditure. The institute, in its analysis of union government budgets, developed a model for a gender wise allocation of government expenditure. However, the areas where benefit to any gender group is not clearly visible, there arises the problem. Assuming this, the claim of the gender budget is to the tune of the 30 per cent of total allocation rather than 50 per cent which is their due if population basis of division is made. This means government schemes are required to be clearly divided into three distinct and mutually exclusive groups. These are women-benefitting schemes (WBS), gender-neutral schemes (GNS), and male benefiting schemes (MBS). As substantial part of benefit of GNS and MBS are likely to be appropriated by the male members, it is necessary to guard for at least 30 per cent of the total allocation to flow to the female members of the society.

R. Sharp (1991) an Australian economist suggested to break the allocation for each department into three categories. These are: (i) expenditure specifically targeted for women; (ii) expenditure promoting equal employment opportunity within the public sector; and (iii) other main stream budgetary expenditure .The assumption in the methodology is that all mainstream budgetary expenditure has gender impacts. The deficiency here is that all schemes and programmes of the government cannot be gender portioned and mutually exclusive.

A. Women Specific Programmes (WSP)

Public expenditures, which are specifically targeted to women and girls, are included in this category. The expenditures incurred in this category can be categorized under the following heads:

1. Expenditures made for some protective and welfare services that are important to prevent the atrocities against women;

2. The expenditure incurred to empower women where they can play their rightful role in the economy. For example, expenditures on education, health and nutrition, housing etc.;

3. Several expenditures, which enable this section for specific self-employment and enhances their capacity through training etc. In other words, expenditures made for economic empowerment of the women includes in this category;

4. Lastly, the expenditures incurred for certain regulatory services and awareness generation programmes for women are demarcated as specially targeted expenditure on women. For example, expenditures made for creating awareness on maternity benefit schemes and other benefit schemes for women.

B. Public Expenditure with Pro-Women Allocation

There are certain other schemes, which are not exclusively targeted for women but have substantial pro-women character. For instance, programmes relating to alleviation of poverty and generating employment, expenditures incurred may have sizeable benefit for women. Public provision of drinking water supply and sanitation, fuel, housing etc. are included in this category. To calculate the share of women's component of composite public expenditure in total expenditure the following formulae are used.

I. The calculations are to be done strictly for illustrative purposes. In the lines of guidance of Women and Child Development Department on the status of women component plan to various department, some departments demarcated the percentage share of women component for certain schemes/programmes, where women constitute a significant part of the beneficiaries of almost all the schemes, is calculated by using the formula given below:

Pro-women Allocation= [TE-WSP] * WC

Where, TE= Total Expenditure

WSP= Women Specific Programmes of the concerned Department

WC= Women Component specified as a percentage of the total outlay of the department being exclusively spent on women.

II. In the second instance, there are several departments under which only certain programmes (not all) are included as women's component. Pro-women allocations of those departments are calculated on the basis of following formula:

Pro-women Allocation =$\sum$ [SCS * WC]

Where, SCS= Expenditure on the specific composite scheme

WC= Women's Component specified as a percentage of the total outlay on the specific scheme (For example 30 per cent stipulation is required in some employment oriented schemes).

First, the schemes with women's component and their pro-women allocations are to be identified and then individual shares should be added together to get the total pro-women allocation of the department/Ministry. For the departments under which the information about women component plan is not provided in the DWCD document, the women component would assume to be 30 per cent in conformity with the objective of Ninth Plan.

C. Benefit-incidence Analysis

Public expenditure on certain area specific schemes/ programmes and community specific schemes where gender desegregation is not possible fall under this category of analysis. Gender disaggregated benefit incidence analysis of public spending reveals the distributional impact of budgetary policies. For calculating such public expenditure by sex, requires the estimates of the cost of providing public services and data on gender beneficiaries. In other words, gender-disaggregated benefit incidence calculation involves the measurement of unit cost of providing a particular service and the number of units utilized by gender. The benefit incidence of public spending on women can be calculated by multiplying the unit cost of the particular service and the number of female beneficiaries. It is obvious that nature of public expenditure on certain public goods and services cannot be gender portioned. The formula for this may be as follows:

Allocation to Women=

Where F = percentage of benefit going to wómen

X = expenditure in the scheme

Critical Concerns of Gender Budget Analysis

Gender budgeting is one of the important tools to deal with gender related problems in the country. It is an innovative approach where the programmes and schemes have allocations in line with the gender. The methodology adopted so far provides a way to analyze the pattern of allocations in case of programmes and schemes exclusively meant for women and pro women schemes like public provision of drinking water supply and sanitation, fuel, housing etc. This apart there are a lot of schemes like mega power projects, dams, irrigation projects and other community development programmes in which it is difficult to find out allocations on gender lines. Even the recently announced Bharat Nirman comes in this category. In such cases it is difficult to identify women and men beneficiaries and accordingly track the incidence of public expenditure. So here the question is how to maintain a gender disaggregated data base at the beneficiary level.

Secondly, huge amount of total expenditure at the state level is non plan expenditure which is utilized for salary and other administrative heads. Due to rise in such expenditure the states are left with a little resource to introduce developmental programmes. So can we disaggregate this expenditure on gender lines? It is also true that many externally aided projects are run at the state level which is not routed through the state budget. It is difficult to find out gender differentials in case of such expenditures.

The most important point here is not the allocation of resources whether the resources reach the targeted beneficiary and how to measure the incidence of public expenditure and its efficiency. A rising trend in public expenditure does not translate into enhanced benefits for women.

Conclusion

In spite of all these difficulties, action in this area is likely to have substantial impact on the growth prospects of the economy resulting in greater women labour force participation. In the future years, the advantage of the Indian economy is its cheap labour force and by use of this it is possible to compete with the so called developed nations effectively. The participation of women folk in certain non-traditional activities like armed forces, space research etc is definitely gives an optimistic future of gender budgeting exercise.

REFERENCES

Banerjee N. and Roy P., (2004), "Gender in Fiscal Policies the Case of West Bengal", Follow the Money Series, UNIFEM South Asian Regional . Office, New Delhi.

Banarjee N., What is Gender Budgeting? , Sachetan, Kolkata.

Debbie B., Diane E., Guy H. and Tanni M., (Jan., 2002), Commonwealth Secretariat, United Kingdom.

Due J.F and Friedlander A.F, (1973),"Government Finance Economics of Public Sector", Richard D. Irwin, Inc.

Gender Budgeting In India (2003), Follow the Money Series, UNIFEM South Asian Regional Office, New Delhi.

Government of India (2001) Census of India: Provisional Population Totals, Series-22: Orissa, Directorate of Census Operations, Orissa.

Katrin H., (Feb, 2004), Gender Budgeting An Overview by the European Women's Lobby, Brussels.

Musgrave, R.A., (1961)," The Theory of Public Finance", Mc Graw Hill, Roga Kusa Ltd., New York.

Sharp R., (July, 2003), Budgeting for Equity: Gender Budget Initiatives within a Framework of Performance Oriented Budgeting, UNIFEM, United Nations, New York.

6

Action for Gender Equality and Women Empowerment by UNIFEM

with Special Reference to India

Mr. Chandra Sekhar Bahinipati and
Dr. Dillip Kumar Mishra

Introduction

UNIFEM-United Nations Development Fund for Women is organized by United Nation for solving every problem regarding women. Following a First World Conference on Women in 1975, the UN General Assembly established the United Nations Development Fund for Women in 1976 through its resolution 31/133. It provides financial and technical assistance to innovative programmes and strategies that promote women's human rights, political participation and economic security. Within the UN system, it promotes gender equality and links women issues and concerns to national, regional and global agendas for fostering collaboration and providing technical expertise on gender mainstreaming and women's empowerment strategies. Since then, it has provided financial and technical assistance to thousands of innovative approaches throughout the world aimed at fostering women's empowerment and gender equality.

This chapter confines to the gender equality and women empowerment by UNIFEM with special reference to India. In the world as well as in India, there are two categories of people living, i.e., men and women. Always women are behind men who are given lower status in the society. In the case of women there is some strict social norms, which is not the case for men. Now, see the developed countries, what happens there in case of women that women are doing better than men and stay at the forefront men. In social, political and economic spheres women are doing better than men. In research and also in defence, women are getting opportunity and doing better. In developing and also under-developed countries, equality and empowerment women are not given full right in all aspects. For that reason, these countries are not grown much as compared to the developed countries.

In India also, women are not given full freedom and empowerment. The country is giving more emphasis on the gender equality and empowerment of women. It must also give right to women in every sphere and extended gender reservation in all, field, which much help women and offer opportunity to show their potential. There are various examples of women outstanding men in some aspects. If women are given a chance then they can show their quality which would help the country and also the society. In India , more than 22 per cent of people are living below poverty line and only male are getting chance for working, which is not sufficient to maintain livelihood of the society. It, women join their hands with men, then household income increases besides increasing national income and productive capacity of the country. For this reason, government and also various NGOs should work to solve their problems. Among these an international organization also works with this agenda i.e. UNIFEM.

Today, the organization's work touches the lives of women and girls in more than 100 countries. It also helps make the voices of women heard at the United Nations—to highlight critical issues and advocate for the implementation of commitments made to women by the world's nations.It is the women's fund at the United Nations, which provides financial and technical assistance to innovative programmes and strategies to foster women's empowerment and gender equality. Placing the advancement of women's human rights at the centre of all of its efforts, it focuses its activities on four strategic areas:

A. Reducing feminized poverty;

B. Ending violence against women;

C. Reversing the spread of HIV / AIDS among women and girls; and

D. Achieving gender equality in democratic governance in times of peace as well as war.

To pursue these goals, it is active in all regions and at different levels. It works with countries to formulate and implement laws and policies to eliminate gender discrimination and promote gender quality in such areas as land and inheritance rights, decent works for women and ending violence against women. It also aims to transform institutions to make them more accountable to gender equality and women's rights, to strengthen the capacity and voice of women rights and advocates, and to change harmful and discriminatory practices in society.

Now UNIFEM orgnise 30th anniversary and on this ceremony, we celebrate the partnerships, leadership and vision of scores of individuals, groups and governments that have made UNIFEM a strong and effective women's fund.On the 30th Anniversary, Noeleen Heyzer, Executive Director, UNIFEM told as, "my wish is for a strong UNIFEM to reach more people with the power to change the conditions under which women work and live, making the world a more just and equitable and a happier place for all". Below, we describe the various views of diffreent people about the prosperity of women.

Elsie Onubogu said, "For me, women's progress is when every woman can make and contribute to informed decisions about her rights, welfare, and general well-being of her society."

Wang Jiax'iang said, "The feminist movement and the demands of women in any particular country grow out of the reality of that country, and it is wrong to say that what we want is what everybody should want and what we do not want nobody should ask for".

Amartya Sen said, "The insecure sharecropper, the exploited landless labourer, the overworked domestic servant, the subordinate housewife, may all come to terms with their predicament in such a way that grievance and discontent are submerged in cheerful endurance by the necessity of uneventful survival. The hopeless underdog loses the courage to desire a better deal and learns to take pleasure in small mercies".

Marilyn Waring mentioned that, "Every time I see a mother with an infant, I know I am seeing a women at work. I know that work is not leisure and it is not sleep and it may well be enjoyable. I know that money payment is not necessary for work to be done. But, again, I seem to be at odds with economics as a discipline, because when work becomes a concept in institutionalized economics, payment enters the picture .... No housewives, according to this economic definition, are workers."

Ajit Singh and Ann Zammit said that, "The Korean government promoted a national slogan 'Get Your Husband Energized' that called on women to help offset the impact of the crisis on men, who on becoming unemployed or bankrupt were subject to depression."

Objectives

The chapter focuses on action for gender equality and women empowerment in India. The main aim of the study is related to the gender equality and women empowerment and how UNIFEM maintain this in the world. The important objectives of the study are as follows:

1. Describe the function and working of the UNIFEM;
2. How UNIFEM functions to bring gender equality and women empowerment in world as well as in India;
3. Derive and critically examine gender equality and women empowerment in India.
4. Examine the conclusion and some policy suggestions.

Functions of UNIFEM

UNIFEM was the internationally reputed organization, working for women in particular. It takes a holistic approach—linking normative and legal frameworks with institutional reform to bring concrete change to women on the ground. This approach brings together efforts to:

1. Formulate and implement laws and policies to promote gender equality and women's human rights;

2. Build institutional capacity to allocate resources and establish accountability mechanisms to ensure implementation;
3. Strengthen gender equality advocates to monitor and track progress and mobilize constituencies to bring about change; and
4. End harmful practices and attitudes that perpetuate gender inequality around the world.

UNIFEM in Relation to Women in World

The important aim of UNIFEM is to reduce feminized poverty and strengthen women's economic security. The Fund concentrated on linking economic decision-making more closely to economic and social development, incorporating gender into poverty reduction strategies in 18 countries, mainly through gender-responsive budget initiatives. These programmes, which apply gender analysis to government budgets to highlight the impact of tax policy and revenue allocations on women and men, show how UNIFEM has brought the power of partnerships, resources and ideas to make a difference—building technical capacity and mobilizing knowledge networks to ensure allocation of adequate financial resources for gender equality and women's empowerment. In 2001, we co-organized a High Level Conference which brought together governments, civil society and gender budget experts to launch a campaign for gender responsive budget initiatives to be undertaken in 45 countries by 2015. Today, we can track results through programmes we have supported in 30 countries.

The power of mobilizing new partnerships for change can also be seen in our work with the private sector to increase women's access to new employment and livelihood opportunities in the global economy. In 2005 we expanded a partnership with Cisco Systems in Arab States to enhance women's knowledge of and access to ICTs (Information and Communication Technologies), including the creation of e-villages in rural areas—and established a new partnership with Macy's Department Stores to open up new global markets for women weavers in post-conflict

Rwanda. In Asia Pacific, our support to governments and women's groups working to improve conditions for women migrant workers resulted in the adoption of a Code of Ethical Conduct for service recruitment agencies in South and Southeast Asia that protects the rights of women who migrate for work in the Arab region.

UNIFEM's work on gender equality in democratic governance and post-conflict reconstruction focuses on building partnerships that can operate as change agents at different levels to promote women's rights and strengthen their participation as voters, candidates and decision-makers. In 2005, it built partnerships with women's groups, civil society, international donors, regional organizations and the media to focus on women's inclusion in all aspects of political participation, peace building and governance. Drawing on our experience in over 20 countries, supported by the Government of the United Kingdom, it has developed an approach that consistently yields results for women in post-conflict countries. In Sudan, for example, it partnered with the Government of Norway to create opportunities for women from the North and South to come together to identify priorities and develop a common agenda for women's rights in post-conflict reconstruction, and present this to the donors' conference in Oslo. This helped to highlight their perspectives and effectively position them to take part in shaping interim constitutions. Article 15 of the interim national constitution prohibits forced marriage and mandates the State to "emancipate women from injustice, promote gender equality and encourage the role of women in family and public life." For the next step, turning these principles into enforceable rights, it is partnering with the African Union and the Inter-Governmental Authority on Development (IGAD) to support national machineries in the North and the South to undertake a comprehensive gender analysis of their interim constitutions as a basis for legal reform and its implementation. Countries emerging from conflict offer a unique opportunity to put in place a gender justice agenda that addresses the implementation of women's human rights by revising laws that discriminate against women, such as inheritance law, personal status law, labour law, and empowering women to access rule of law and economic and social justice institutions.

In the area of ending violence against women, it recognizes that addressing the multiple forms of violence that women face, both during conflict and in times of peace, is a long and difficult process. It requires strategies to engage men in transforming power relationships and coordinated strategies at multiple levels and across sectors to address the social and economic causes of gender violence and the links between violence, poverty and HIV/AIDS. In Latin America and the Caribbean, within the highly visible 16 Days of Activism Campaign, UNIFEM leads an annual coordinated UN Interagency Campaign against Violence against Women, bringing together governments, women's rights and civil society activists and the media to bring coordinated attention and response to this issue as human rights violation and public health crisis. We are also working in three Latin American cities on a Safe Cities for Women Project to address problems of women on the ground with regard to security and ensure they are included in urban security decision-making processes.

The United Nations Trust Fund to Eliminate Violence against Women, established in UNIFEM by General Assembly resolution 50/166, now focuses consistently on securing on-the-ground implementation of the laws and policies to address this issue that are now in place in 89 countries. Trust Fund strategies include campaigns to raise awareness of existing laws and to bring these into conformity with human rights principles; the analysis of budgets and allocations needed to support integrated efforts to implement laws; capacity building for judicial, law enforcement, health and other officials to comply with laws and policies; creation of women friendly law enforcement systems that enable women to claim their rights; and creation of data collection systems and indicators to help monitor the extent to which laws and policies are being implemented to reduce violence against women. In a new partnership with the Global Coalition on Women and AIDS, the Trust Fund is specifically addressing the links between gender violence and HIV/AIDS—for which it has secured, for the first time, private sector support of nearly $1 million with the expectation of another $1.4 million if performance matches its expected potential. It also works to combat the spread of HIV/AIDS through increased attention to the gender dimensions of this pandemic, partnering with national AIDS councils, UNCTs, gender equality advocates and women living with HIV/AIDS to press for legislative reform related to gender inequality and HIV/AIDS policies, and with national and local institutions to adopt gender-responsive policies and practices to make a positive difference in women's lives. In India, for example, our partnership with the National AIDS Control Organization (NACO) builds their capacity to use analysis and information to break through stigma and silence to make the links between gender-based violence and HIV/AIDS.

UNIFEM's work on CEDAW (Convention on the Elimination of All Forms of Discrimination against Women) implementation illustrates the importance of building sustainable partnerships at national and local levels to use this powerful instrument to bring about change—through stronger legal and policy frameworks for gender equality and stronger mechanisms for implementation, monitoring and reporting. With support from the Government of Canada, it works with government institutions and women's groups in over 20 countries to implement CEDAW, including through regional programmes in Southeast Asia, Arab States and Pacific Islands. The examples I have highlighted show that to deliver on gender equality, the issue is not lack of good practices even lack of know-how. We know how change happens. The issue is how to implement strategies and practices on a scale that is large enough to turn the tide for gender equality and women's rights. We are at a crossroads in our work on gender equality. We have worked hard to secure commitments and mechanisms, even models for change. We can continue on this road or take the more challenging road of investing in what has delivered for women and changing what we know is not working. Today, in the context of UN reform, renewed commitment to achieving the Millennium Development Goals (MDGs), and the rollout of the new aid effectiveness agenda, there is an unprecedented opportunity to expand effective strategies and practices that have worked to assist countries to deliver on gender equality and women's empowerment and achieve the MDGs.

In line with harmonized efforts to reach the target for goal , halving the number of people living in poverty by 2015, for example, UNIFEM is implementing an innovative initiatives component of the World Bank's Gender Innovation Fund, so that innovative approaches to enhancing women's economic security can be scaled up into mainstream macro-economic policy. It is also working closely with the Bank to develop an approach to evaluating community-based programmes supported by the Trust Fund that will enable the Bank to build its internal case for addressing the need to end violence against women in its grant and loan operations. However, despite the global consensus that gender equality and women's empowerment are essential to achieving the MDGs, there are as yet no strengthened institutional structures, increased resources or effective accountability and monitoring mechanisms to assist countries to advance gender equality. The sole MDG target for 2005, gender parity in primary and secondary education, has already been missed. Women are paying the costs of this institutional neglect in lives and lost opportunities.

To harness the power of women change agents and begin to turn the tide on gender equality and women's rights, let me highlight three priorities:

1. Strengthening gender equality in national development strategies. Over the last ten years there have been enormous improvements in the normative environment for gender equality. Action plans for gender equality are in place in over 120 countries, laws and action plans to end violence against women have been passed in 89 countries. What is lagging is implementation. Governments are now coming to UNIFEM for assistance in next steps. One of the responses we have made is to provide technical support to national partners to adopt harmonized gender equality indicators for monitoring implementation of the Beijing Platform for Action, CEDAW, and strategies for achieving the MDGs. We are working to strengthen capacity among stakeholders to incorporate harmonized indicators and sex-disaggregated data into national development planning and poverty reduction strategies. In the context of the aid effectiveness agenda,

together with the European Union (EU), we have mobilized a critical mass of change agents committed to the vision of a gender-equitable society. They include leaders at all levels of government who control critical levers for change and financial resources as well as international and regional organizations that provide support to both national governments and civil society organizations. In so doing, we are working to ensure that national ownership means bringing women's voices and perspectives into the development process, that gender equality principles and strategies are incorporated into all coordination mechanisms, including PRSPs and sector-support strategies, and that national plans of action for gender equality as well as action plans to end violence against women are integrated into national development planning and strategies;

2. Strengthening a coherent and integrated approach across the UN (United Nation) system. A more vigorous effort is needed to ensure that gender equality is addressed in a coherent way throughout the UN system, especially in its operational activities on the ground, where it matters most. The UNDG (United Nation Development Goal) Task Team on Gender Equality, which UNIFEM chairs, coordinates action among its 16 member agencies in mainstreaming gender equality and women's empowerment and ensuring these are incorporated into the tools and guidance the UNDG gives to UN country teams. The Task Team has recommended bringing together good practices in each organization to "raise the bar" on accountability for gender equality across the UN system. It proposes two actions, endorsed by the UNDG Principals, to support UNCTs to move from improved analysis to more coherent implementation and accountability:

   (a) an "Accounting for Gender Equality" scorecard that sets minimum standards for UNCTs to assess their performance and identify gaps and progress across the system; and

(b) an "Action Learning" process with a number of UNCTs to generate strategies to undertake rights-based, change-oriented programming that supports government and civil society to move forward on gender equality and women's empowerment;

3. Strengthening monitoring and accountability by women on the ground. Accountability means different things to different people, depending on where they are located vis a vis the power to bring about change. For gender equality and social justice advocates, accountability means ensuring sufficient resources for implementation as well as expanding community-level participation in defining targets and tracking progress towards them. This not only supports greater inclusion in national planning but also supports organizing by grass-roots and women's organizations to exercise a watchdog function to ensure national resources are allocated all the way to the ground, and bring realities from the ground to inform policy direction. In so doing, it works to bring underrepresented and excluded groups, such as HIV positive women, women informal workers, indigenous women, and rural poor women, into the development process. In the context of UN reform, there is general agreement that the "gender equality architecture" must be strengthened—at both the global and the country level. We know so much more now than we did even ten years ago about supporting gender equality and women's empowerment. What we need is the courage to build on what we know to bring about the radical changes needed to enable the UN system as a whole to respond to the new opportunities for—and current threats to—gender equality and women's rights, providing the kind of policy advice and technical expertise that will help countries to deliver on these goals.

However, a consolidated gender entity that mainly issues "policy advice" or offers catalytic examples will not enable us to fully respond to the opportunities for—or the current threats to—greater gender equality. This entity can only make the much-needed difference if it can inspire and mobilize partnerships and agents for change and if it has sufficient presence, authority and resources at the UN Country Team table. To make this kind of difference it must be connected to a constituency at the country level that can ensure that the system is responsive to the actual lived realities of women, particularly poor and excluded women. Finally, it must operate as a driver that can stay ahead of the curve to take on emerging issues and push the system to take them on. If the UN is to remain a legitimate development player in the 21st century, it must stay at the forefront to assist countries to deliver on gender equality and women's empowerment.

The Effective Agenda of Gender Equality

Gender equality is central to achieving the MDGs and other development goals, asking is important to ensure that aid structures target and monitor progress towards gender equality goals. Ultimately, gender equality outcomes will be important signs of the effectiveness of the new approach to aid delivery and partnership. To support gender equality, the new aid architecture should include:

- Adequate financing for programmes that respond to women's needs;
- Accountability systems for governments and donors to track and enhance their contributions to gender equality; and
- Gender-sensitive progress assessments, performance monitoring and indicators for aid effectiveness.

This note is an outcome of a November 2005 international consultation in Brussels organized by the United Nations Fund for Women and the European Commission. It identifies an initial set of considerations to ensure that gender equality is central to the aid effectiveness agenda. This note is intended for policy-makers currently adjusting to the new aid modalities—such as officials and analysts in Ministries of Finance, Planning and Women's Affairs; women's rights advocates at domestic, regional and international levels; and bilateral and multilateral development actors such as Resident Coordinators in the UN system.

Women's International Partnerships for Change

Women are increasingly organizing through international networks and coalitions which bring diverse women together to negotiate and pursue common objectives. Partnerships with which UNIFEM works include:

- DAWN (Development Alternatives with Women for a New Era), a network of women scholars and activists from the economic South who engage in feminist research and analysis of the global environment and are committed to working for equitable, just and sustainable development.(Website: www.dawn.org.fj)
- GROOTS International (Grassroots Organizations Operating Together in Sisterhood), a global network of women's groups committed to developing a movement giving voice and power for change to low-income and poor women's initiatives. Organizations within the network are active in areas such as credit, asset creation and small business development, sustainable agriculture, food processing, housing, popular education, health and bottom-up community development planning. (Website: www.jtb-servers.com/groots.htm.)
- WIEGO (Women in Informal Employment Globalizing and Organizing), a worldwide coalition of individuals from grassroots organizations, academic institutions and international development agencies concerned with improving conditions for women in the informal economy, through better statistics, research, programmes, and policies. (Website: www.wiego.org)
- Home Net, an international network of women's groups, trade unions and other civil-society

organizations advocating for the rights of home-based workers. Home Net was pivotal in securing NGO and government support for the International Labour Organization (ILO) Convention on Home Work, adopted in 1996, and campaigns to make governments aware of the need to ratify this and all other ILO Conventions.(Website: www.gn.apc.org/homenet.).

Guideline of UNIFEM for Women Empowerment

UNIFEM's guidelines on women's empowerment (1997a) include:

- Acquiring knowledge and understanding of gender relations and ways in which these relations may be changed;
- Developing a sense of self-worth, a belief in one's ability to secure desired changes and the right to control one's life;
- Gaining the ability to generate choices and exercise bargaining power;
- Developing the ability to organize and influence the direction of social change to create a more just social and economic order, nationally and internationally.

Achieving this requires both a process of self empowerment, in which women claim time and space to re-examine their own lives critically and collectively; and the creation of an enabling environment for women's empowerment by other social actors, including other civil-society organizations, governments and international institutions. This concept of women's empowerment goes well beyond women's participation in agendas set by others. It entails both the development of women's own agency and the removal of barriers to the exercise of this agency.

Budgeting for Gender Equality in World and India by UNIFEM

Gender-responsive budget analysis looks beyond the balance sheets to probe whether men and women fare differently under existing revenue and expenditure patterns. This process does not involve creating separate budgets for women, or aim solely to boost spending on women's programmes. Instead, it helps governments understand how they may need to adjust their priorities and reallocate resources to live up to their commitments to achieving gender equality and advancing women's rights—including those stipulated in the Convention on the Elimination of All Forms of Discrimination against Women, the Beijing Platform for Action and the Millennium Development Goals. Engendered budgets can be critical to transforming rhetoric about women's empowerment into concrete reality.

Assessing budgets through a gender lens requires thinking about government finances in a new way. It calls for including equity in budget performance indicators, and examining impact of budget policies on gender equality outcomes. It also focuses on the relation between government spending and women's time spent in unpaid care work such as water and fuel collection, caring for the sick, childcare and many others. Conducting a gender-responsive budget analysis can be seen as a step not only towards accountability to women's rights, but also towards greater public transparency and economic efficiency. With compelling evidence that gender inequality extracts enormous economic and human development costs, shifting fiscal policy to close the gaps yields gains across societies.

Working in close partnership with women's organizations and scholars, UNIFEM has helped pioneer cutting edge work on gender-responsive budgets that is being picked up by both local and national governments. Advocacy and training for government officials, parliamentarians and women's groups, the development of budget analysis tools and widely shared knowledge on what works have helped the concept catch on, resulting in changes in a number of countries.

In India, several years of sustained advocacy and partnerships among the Department of Women and Children, UNIFEM and other women's organizations have encouraged the national Government to affirm the importance of gender budgeting, initially through the inclusion of a gender budget statement in the 2003 Union Budget and through official studies of the issue. In 2005, the Finance Minister committed to moving forward on implementation. Twenty-one national ministries have now set up gender-budgeting cells. For the fiscal year 2005-2006, 18 departments are rolling out detailed specifications of allocations and targets benefiting women. On the state level, in West Bengal, it supported the organization Sachetana to prepare a gender budgeting manual that the group has used to train over 1,000 women councilors in local governments. In Karnataka, another state, elected women representatives in the city of Mysore used gender budgeting to ward off a proposed budget cut targeting women's programmes. They ended up securing a 56 per cent increase in funding instead, and started advocacy for more transparent public information in the future. At the local level, GBIs (Gender Budget Initiatives) in three states in India (Karnataka, West Bengal and Maharashtra) devised innovative approaches for holding local governments accountable to women's priorities. Successful examples were generated in terms of greater accountability of municipal councils and mayors to women's concerns as well as introduction of concrete changes in resources available to women at local levels. In Karnataka, elected local women leaders have advocated successfully with the Chief Accounts Officer of Mysore to double resource allocations to women's priorities and to reintroduce a women's health insurance scheme. In West Bengal, women local councilors are raising questions to the local government on proper recording of agreements reached at Panchayat (local council) meetings and on how beneficiaries of special safety net programmes for those living below the poverty line are identified.

Conclusion

The paper discusses about the action of gender equality and women empowerment by UNIFEM—with a special reference to India. In this paper, we discussed the leading role of the internationally reputed and co-operative institute UNIFEM in relation to the gender equality and women empowerment. It is the international reputed organization of United Nation especially for women, which has always been trying to bring gender equality and women empowerment. It was formed in 1975 and recently it celebrated its 30th anniversary. It makes relationship with more than 100 countries, organize various conferences and provide funds through Government and different NGOs for eradicating or combating the violence against women and put their entire effort to bring gender equality and women empowerment. For providing women's human right at the center, it follows four strategies i.e., (A) Reducing feminized poverty; (B) Ending violence against women; (C) Reversing the spread of HIV/AIDS among women and girls; and (D) Achieving gender equality in democratic governance in times of peace as well as war.

In world as well as in India, most the people face the problem of hunger or poverty and their daily income is less than $1 or $2 per day. Due to the causes of this, most of the women not get a chance for empowerment. For this, UNIFEM work seriously for eradicating poverty and provide her employment opportunity through establishing ICTs (Information Communication Technologies) and e-villages in rural areas. The above function has dual benefit i.e., one, reduction of poverty and other, for providing gender equality and women empowerment.

It established the United Nations Trust Fund to Eliminate Violence against Women which focuses consistently on securing on-the-ground implementation of the laws and policies in 89 countries. Trust Fund strategies include campaigns to raise awareness of existing laws and to bring these into conformity with human rights principles. In a new partnership with the Global Coalition on Women and AIDS, the Trust Fund is specifically addressing the links between gender violence and HIV/AIDS. It also works to combat the spread of HIV/AIDS through increased attention to the gender dimensions of this pandemic, partnering with national AIDS councils, UNCTs, gender equality. It also partnership with the National AIDS Control Organization (NACO) builds their capacity to use analysis and information to break through stigma and silence to make the links between gender-based violence and HIV/AIDS in India.

It formulates CEDAW implementation to bring about gender equality and women empowerment. With support from the Government of Canada, it works with government institutions and women's groups in over 20 countries to implement CEDAW, including through regional programmes in Southeast Asia, Arab

States and Pacific Islands. To achieve MDGs, gender equality and women's rights, it focused three properties: (1) Strengthening gender equality in national development strategies; (2) Strengthening a coherent and integrated approach across the UN (United Nation) system; (3) Strengthening monitoring and accountability by women on the ground. For the fulfillment of the common objectives of women empowerment and gender equality, it perform works i.e., DAWN, GROOTS International , WIEGO and Home Net.

In India, the Department of Women and Children, UNIFEM and other women's organizations have encouraged the national Government to prepare the gender budgeting. In 2005, the Finance Minister implemented this and now twenty-one national ministries have now set up gender-budgeting cells. In West Bengal, it supported the organization Sachetana to prepare a gender budgeting and in Karnataka elected women representatives in the city of Mysore used gender budgeting. At the local level, Gender Budget Initiatives (GBIs) in three states in India (Karnataka, West Bengal and Maharashtra) devised innovative approaches for holding local governments accountable to women's priorities. UNDP-UNIFEM endeavor to incorporate gender concerns into the Eleventh Five Year Plan, took place in April 2006 in Bangalore, drawing a hundred participants from civil society and government. Sessions were held to discuss the issues of Dalits and adivasis, minorities, especially Muslims, the urban poor, unorganized sector workers, women in local self-government and women in disaster situations, especially tsunami affected women. All the participants agreed that a specific mechanism at the State Government Level is needed to ensure that local and district level plans reach the Planning Commission. They also called for a gender auditing and monitoring mechanism to look into the impact of the plan on women. They focused on land and agriculture issues and recommending that women's concern be made more visible in agricultural policy.

UNIFEM and UNDP are working together in collaboration with the Ministry of Women and Child Development, the Planning Commission of India and decide to make group of experts to incorporate gender concerns in Eleventh Five Year Plan period.

The key points on this discussion include: ensuring property right for women, establishing social safety nets, increasing allocation for women health programmes and building the capacity of health workers, implementing Domestic Violence Act and strengthening support services for victims of violence against women (VAW), setting of community care homes for HIV-positive women, making provisions for services benefiting women working in the informal sector and domestic services and provide support to increase number of women judges, doctors and teachers.

The whole paper intends the issues of women empowerment and gender equality by UNIFEM with a special reference to India. It also discusses the entire function and the policy of the UNIFEM and their working in different countries and also in India At last make some policy formulation with the Government of India for enhancing the programme of women empowerment and gender equality.

REFERENCES

A UNIFEM Briefing Paper,"Human Rights Protections Applicable to Women Migrant Workers", UNIFEM Regional Programme on Empowering Women Migrant Workers in Asia.

DAC Working Party on Gender Equality, 2001, 'Report on the Consultation Workshop on Gender Equality in Sector Wide Approaches', The Hague, 22-23 February.

Driscoll K. and D. Booth, 2005, 'Progress Reviews and Performance Assessment in Poverty-reduction Strategies and Budget Support: A Survey of Current Thinking and Practice', ODI, UK, May http://www.odi.org.uk/publications/ reports/JICA_report_web.pdf.

En Route To Equality: A Gender Review of National MDG Reports', August 2005, http://www.undp. org/ gender/docs/en-route-to-equality.pdf.

Heyzer, N.,"Turning the Tide:The Next Wave of Action for Gender Equality and Women's Empowerment",61st Session of the UN General Assembly, Third Committee Agenda Item 60 (a): Advancement of Women 9 October 2006.

http://www.oecd.org/dataoecd/24/17/1956174.pdf

http://www.oecd.org/dataoecd/45/46/35230756.pdf.

http://www.un.org/esa/sustdev/documents/Monterrey_Cons ensus.htm.

http://www.unifem.org/attachments/products/153_chap1.pdf

http://www.unifem.org/attachments/products/154_chap2.pdf

http://www.unifem.org/attachments/products/155_chap3.pdf

http://www.unifem.org/attachments/products/156_chap4.pdf

http://www.unifem.org/attachments/products/157_chap5.pdf

http://www.unifem.org/attachments/products/158_chap6.pdf

http://www.unifem.org/news_events/event_detail.php?Event ID=31.

http://www.unifem.org/news_events/event_detail.php?Event ID=31.

ILO–GENPROM Working Paper No. 10: Rima Sabban, Migrant Women in the United Arab Emirates: The Case of Female Domestic Workers, at 20.

Mpagi, Sanyu Jane, 2005, 'The National Development Policy Process—Opportunities for Engagement: A Case of Uganda', Paper Prepared for the UNIFEM—EC Consultation: *Owning Development: Promoting Gender Equality in New Aid Modalities and Partnerships*, 9-11 November, Brussels, Belgium.

Painter, G, 2004, 'Gender, the Millennium Development Goals and Human Rights in the Context of the 2005 Review Process', Draft Report for GAD Network, Commissioned DFID.

Roundtable on the Human Rights of Women Migrant Workers Under CEDAW 11 July 2003.

Tendler, J., and S. Freedheim, 1994, 'Trust in a Rent Seeking Society: Health and Government Transformed in North-eastern Brazil', *World Development* 22 (12).

UNIFEM Discussion Paper–March 2006,"Promoting Gender Equality in New Aid Modalities and Partnerships".

Van Diesen, A. and J. Yates, 2005, 'Gender Mainstreaming in the Context of Poverty Reduction Strategies and Budget Support—The Case of Uganda', mimeo.

www.dawn.org.fj

www.gn.apc.org/homenet.

www.jtb-servers.com/groots.htm.

www.unifem.org

www.wiego.org

7

Violence Against the Women and Gender Justice

Mr. Deepak Bishoyi

Human beings are born equal but are never treated equal. Unequal treatment to women or the gender bias is not a new phenomenon. Our mythology, if referred to shows us the gender bias that was prevalent in India. May it be Sita or Draupadi both had to bow down before this biased treatment. Even to-day women continue to be victims. Since the idea to celebrate Women's Day took shape at the second International Conference of Socialist Women in Copenhagen sometime back in 1910, the world has definitely seen some change in the status of women. With the changing environment and the changing world around her, the women have broken the shackles of confinement and has chosen to play an important role in decision making, be at home or elsewhere. She fought for her rights, for a place in the still-patriarchal society. Even though women have entered the assembly and parliament not much has changed in the status of women and not much have come to them. The general condition of women has not changed. Gender bias is still there. Gender bias is the cause of high acquittals in crime against women. Their husbands are abusing every day countless women psychologically, physically and sexually. Despite the feminist movement gaining ground, women continue to be battered, bruised and humiliated in our patriarchal society. However it is not legislation that is lacking but the will to interpret it in favour of women.

Gender bias among law enforcing agencies especially police and judiciary was for a very high rate of acquittal in cases of crime against women. The bigamy law makes it impossibie to marry twice anywhere in the world, but in India the people are openly violating it as under the law, only an affected wife could complain against the husband. Also under the law a man can sue adulterous wife and not vice versa. The gender bias has two aspects. The condition of rural women is more pathetic as they do not enjoy much of rights and are treated shabbily by the law enforcing agencies. Therefore, the position of women needs to be taken seriously to improve the attitude of law enforcing agencies. Hence the investigating wing of the police needs to be separated from its law and order maintenance duty so that violence on women are brought to book in right earnest.

Crime against women beings even before she is born and continues till she breaths her last. Today a girl child is not even allowed to be born. Female foeticide has become a normal issue. After she is born throughout her childhood she has to adjust to the discrimination being made by her parents between she and her brothers. During the teenage she has to digest whatever is meted out. She has to satisfy her family, attend on her household work and her children, should be in time at the work spot and come to the expectations of the boss.

At the time of marriage a girl is considered no better than a commodity. In many cases a girl is not even enquired about her opinion regarding the groom. After the marriage the lineage of the girl is charged and literally she loses her identity. Hence, onwards she has to be submissive but if she differs with her husband or in laws she is branded as arrogant. Further, unfortunately if without any fault of her she is divorced she will not be accepted whole-heartedly either by her parents by her own children because she becomes weak and cannot serve them any more as earlier.

Crime against women like, molestation, bride burning, dowry deaths have become so common that a news paper cannot be completed without such news. The life of some girl is not secured even at home. Father, father-in-law, brother,

brother-in-law, uncle and neighbour in many cases make the life of the girl a nightmare. The male community (not 100%) has become so merciless that an infant also cannot be assured security.

Global Scenario

Gender based violence has emerged as a global issue crossing the regional, social, cultural and economic boundaries of the countries. It threatens the well-being, rights and dignity of women (Fischback, 1997). According to state statistics, about 18 per cent of women are being sexually abused in the US. As per the UN report the other developed countries like Denmark, Germany, UK, Switzerland could not provide the accurate statistics. In US the department of justice reported that, every year 3-4 million women are battered by their husbands or partners. Six well designed studies from the US suggested that between one in five and one in seven, US women will be the victims of a completed rape in her lifetime (KOSS, 1993). Studies of rape in other parts of the world, studies of rape among college age women in Canada, New Zeeland, the UK and the US reveal remarkably similar rates of completed rape across the countries. The studies conducted by Rallies, Watts and ZW (1994) reveals that in developing countries they are in wide spread prevalence of physical and sexual abuse on women (ranging from 12% to 75%) by present or former partners. A study of 769 women in Japan revealed that 59 per cent suffered physical abuse by partners, 66 per cent reported emotional abuse and 60 per cent reported sexual abuse (Domestic Violence Research Group 1993). As far as the African states are concerned the studies made in Kenya, Uganda and Tanzania indicated that from a sample of 733 women from Kissi district in Kenya (Contraceptive prevalence survey) 42 per cent were beaten regularly, 46 per cent of women are physically abused by their partners and 60 per cent in Tanzania had been physically abused. More then 50 per cent of married women studied from Bangkok's biggest slum and construction sites were beaten regularly by their husbands. In China, XU (1997) pointed out that 57 per cent of the wives were reported to be abused by their husbands.

Indian Scenario

The Indian Government while presenting its country report in the 4th World Conference on Women at Beijing in 1995 had recognized violence against women as one of the eleven critical areas of concern. Most of the data are taken from small ad-hoc studies and India is yet to conduct a national survey in different aspects of violence against women. As per the Beijing Platform for Action, violence against women is defined as "an act of gender based violence that results in, or is likely to result in physical, sexual or psychological harms or suffering to women, including threates of such acts, coercion or arbitrary deprivation of liberty whether occurring in public or private life.

The Beijing Platform contained 3 basic objectives, none of which has been implemented by the Indian Government.

(a) To take integrated measures to prevent and eliminate violence against women.

(b) To study the causes and consequences of violence against women and effectiveness of preventive measures.

(c) To eliminate trafficking in women and assist victims of prostitution and trafficking.

When Usha Dhiman of U.P was stripped outside the courthouse of Sharampur after a dispute with her Jot neighbours, no one came to her rescue. The police instead of noting her complaint, as she stood naked asking a police man for a piece of cloth to cover herself, took her into custody and allegedly misbehaved with her. While the local administration took no note of the report of the crime in the locality, small town newspapers, the authorities in Delhi did not even know about it. The credit for the spontaneous follow up has to be given to the Prime Minister and to the National Commission for Women. The chairperson of the Commission (at that time in 1994) Smt. Jayanti Pattnaik rushed to Nayagaon to investigate. Then Dhimam and her husband were asked to come to Delhi to testify before the Commission. The Commission's investigations and report have been held back since the state Government has instituted criminal proceedings in the matter. Uttar Pradesh Chief Secretary has been periodically keeping the Commission informed. The SSP Sharanpur was transferred out of the district, the inspector of the Sadar Bazar Police station and station officer of the Rampur police station was suspended. The amendments to the IPC, the grassroots Mahila Mandals and the growing Women's movement in the country have not been able to check this assault on women's privacy. There is not even the record of these incidents of stripping in the National Commission for the Scheduled Castes and Scheduled Tribes or the National Commission for Women (NCW). Though the Women's Commission has been faithfully following up whatever cases are reported to it, they can only investigate and forward their findings to the state government for action. Police officers are suspended, transferred when the heat from Delhi increased. Then six months to a year later they are in their beat. Caste policies and political interferences make delivery of justice even more difficult.

In 2001 the Institute of Social Sciences organized a workshop on "Gender bias in law enforcement". The activities stressed the need for proper training and sensitization of police to make it more humane. Speaking on the occasion justices Sujata V. Monohor, member of the National Human Rights Commission said attitude of police towards cases of violence against women was normally unsympathetic perfunctory. Police does not pay adequate attention to cases of domestic violence, rape or dowry death. False and perfunctory medical reports are often produced in case of rape while there is a tendency to treat the dowry death as a natural death in accident or a suicide. Seventy per cent of dowry death cases are recorded as kitchen accidents by a stove burst though the stove never is seized. Cases of violence against women were often not investigated properly, sometimes the evidence is destroyed. The women cells too is rendering relief to women in such distress besides the language used by the police is often full of slang and unparlimamentary words which deter women from approaching the police and seek justice.

Report of violence against women, physical as well as psychological, have become a common thing today. Rape, dowry deaths, witch hunting are not just starry incidents but may occur to people from any walks and station of life. In a state like Orissa where the crime graph shows a steady rise, there is strong need to sensitize not just the common mass but police personnel, members of the judiciary and the bureaucracy. This idea was the underlying objective behind setting up of the Task Force on Women and Violence (TFWV) a forum, which aims at sensitizing the society by public education campaigns and legal and structural reform activities. While non-governmental organizations progressed in the right direction did work towards the uplift of women, they failed in gaining an insight into the mechanism, which triggers sexual violence. But the task force set out to do just that the forum established in 1994 by a few members from different organizations, initially worked as a temporary unit with no office. Oxfam supported the initiative. An ignored the dynamics behind crime against women as it opened up a Pandora's box. It was the basic fear to lode into our patriarchal society. The primary areas, which TEWV focused on initially, was on the declining female sex ratio in Orissa, and on the need to sensitize people towards the trauma of a women victims and to campaign for the registration of marriages.

Violence against women is the most pervasive violation of human right in the world today. Its forms are both subtle and its impact on development profound. The specific difficulties that women face relates to their extraordinary work burden or lack of access to health care, for instance do not arise out of ecological deterioration at home and outside but by the specific nature of the task they do (Kusum, 2001).

Needless to mention that human beings routinely subjected to assault, rape sexual slavery, arbitrary imprisonment, torture, verbal abuse, mutilation, ever-murder-all because women were born in a particular group. Their sufferings were compounded by systematic discrimination and humiliation in home and work place, in class rooms and court rooms, at worship and at play, at public places and private surroundings.

However, opening the door on the subject matter against the women is like standing at the thresholds of an immense dark chamber vibration with collective anguish, but with the sounds of protest throttled back to a murmur. For terms of millions of women to day, home is a locus of terror. It is not the assault of stranger that women need fear the most, but everyday brutality at the hands of relatives, friends and lovers. Battering at home constitutes by far the most universal form of violence against women and is a significant cause of injury for the most universal form of violence against women and is a significant cause of injury for women of reproductive age. Yet it is not the sort of act that commands headlines because it happens behind closed doors and because victims fear speaking out. Indeed, domestic violence is tragically common occurrence. It occurs across education, class, income and ethnic boundaries (Chatterji, 1990).

To know about the material facts of violence against the women in India to day, the following top heads are reflected in this manner.

- "Dowry death" is a real fact matters violence against women in India, women is killed because she is unable to meet her in-laws demands for dowry. Over a dozen women a day die as a result inadequate of dowries. Mostly kitchen fires are designed to look like accidents.
- Son's preference is another insidious force directed against women in India.
- Sexual harassment has been publicly acknowledged as harmful to women, it paved the way for prostitution practice.
- Throwing acid to disfigure a woman's empowerment, at present in the transitional phase. Women are not adequately empowered in practice to play an effective role because their male representatives dominate them.
- Forced marriage is a major cause for violence against women in India.
- Victims of rape is most important from of violence against the women in India today.

The age wise break up victims of rape upon women is revealed in the following table.

Table 7.1: Victims of rape upon women, 2000 (age wise)

India/States	16-18	19-30	31-50
Total State	4,614	6,626	1,993
Total Union Territory	08	12	01
India	4,622	6,638	1,994

Source: National Crime Records Bureau, Ministry of Home Affairs, (2002), New Delhi.

In India poverty, discrimination, ignorance and social unrest are common predictors of violence against women. Yet, the most enduring economics of a woman's dignity and security are cultural force aimed at preserving male dominance and female subjugation often defended in the name of venerable tradition.

Stopping violence against is not just a matter of punishing individual acts. It is only when women are strong and equal members of society that violence against them will be as a shocking aberration rather than an invisible norm.

Towards the Gender Justice

The constitution of India guarantees democracy, socialism, secularism, justice, liberty, equality, fraternity and aims at the elimination of social inequalities, economic disparities and political privileges. Article 15 of the constitution states that "the state shall not discriminate any citizen on the grounds only of sex". Further it says, "for the full development of our human resources, improvement of women and building the character of children during the most impressionable years of their infancy, the education of girls is of great importance than that of boys. Article-14 prohibits discrimination on the ground of sex. Article-15(3) of the constitution permits the state to make special provisions for women and children. Article-39 enjoins the state to provide adequate means of livelihood to men and women. Article-51A(e) makes it a duty of every citizen to renounce practices derogatory to the dignity of women. In the light of the constitutional provisions, it is noticed that several forms of gender-biased discrimination still exist. The full potential of women remains under-utilized. The implementation of these laudable constitutional postulates has not been satisfactory. Women constitute about half of the population. They should get their equal rights, opportunities and status with their counterparts. But, still-today, after the 60 years of independent framework, out of fifty per cent of human assets are neglected and are deprived of their human rights for living. Human rights are rights that are held simply by being human being. They are based on the notion of the inherent dignity of every individual. The respect for the inherent dignity of the individual requires that every individual is treated equally and must not be subjected to discrimination. Man, as a member of human society, has some rights in order to survive as well as to make the life better. All human beings possess them by virtue of being members of the great human society irrespective of the fact that they know about them or not.

In the global context, international human right laws are codified in various declarations and conventions to protect the human beings in the spheres where most violation of human rights is taking place. But it is seen that half of out human capital i.e. the women, is being oppressed, harassed and humiliated and treated as commodities and slaves. To improve the status and to protect the human rights of these oppressed women folk United Nation's Women conventions are at work to eliminate the prevailing discriminations against women. March, 8th of every year is being observed as the International Women's Day, to make people aware of the rights of women and to promote the status of women at large. It is felt by great people of India and world that "there is no chance for the welfare of the world unless the condition of women is improved. It is not possible for a bird to fly on one wing".

A woman is raped in India every 30 minutes, only 30 per cent of such incidence gets reported. Only 5 per cent of the rapists brought to trial get convicted. So now the question arises, will death penalty to rapists bring to heel the brutal crime of a man against women?

A number of women activities and the youth believe that the existing law should be implemented stringently without having scope for loopholes to ensure justice for the victims. Though women's right activities, lawyers and NGOs oppose capital punishment for rapists, the younger generation and the policy makers are strongly in its favour, at the same time, they also agree that speed trial is more important for justice to prevail.

With the above background, this paper has made an attempt to study the prevailing gender discriminations and sufferings of women, its impact on the decreasing women population day by day. The paper offers some remedial measures to solve and reduce the issue of gender inequalities in a country like India.

REFERENCES

Bhandare, M. (1994), *The World of Gender Justice*, Har Anand Publication Pvt. Ltd., New Delhi.

Chatterji, A. (1990), "Women in Search of Human Equality", *Social Action*, Vol. 40.

Choudhuri, D. and S.N. Tripathy, (2005), *Girl Child and Human Rights*, Anmol Publication Pvt. Ltd., New Delhi.

Dreze, J. and A.Sen, (1995), *India: Economic Development and Social Opportunity*, New Delhi, Oxford University Press.

Kusum (2001), "How Effective are the Law?", *Yojana*, August.

Saxena, A, (2004), "Women Workers in the Unorganized Sectores: Inequality and Discrimination", Social Action, Vol. 54.

Women's Empowerment, Gender Equality, and the Millennium Development Goals, http://www.un.org/millennium/declaration/ares552e.htm .

UNESCO (1990), *Equality of Access of Women Literacy*, United Nation.

8

Gender Dimension of Internal Migration

A Study of Major States in India

Mr. Ananta Basudev Sahu

The phenomenon of migration in society is universal and is applicable to all stages of human life. Any movement of people from the place of origin to another, involving the permanent change of residence is called as migration. It is one of the most important components of population change. It takes place both internal and external spheres. External migration stands for crossing the international recognized boundaries of the country and settling down in foreign land. The internal migration occurs within the country from one region to another or from one place to another and has psychological, socio-economic and other reasons and backgrounds. In 2000 there were 175 million international migrants in the world, meaning one out of every 35 persons in the world was an international migrant (including both refugees and other international migrants). Based on the world population of 6.057 billion in 2000, migrants represent some 2.9 per cent. It is said that if all the international migrants lived in the same place, it would be the world's fifth biggest country (World Migration 2003). The common perception is that migrants are predominantly male. In fact, global estimates by sex confirm that since 1960 number of female cross-border migrants reached almost the same number as male migrants. By 1960, female migrants accounted for nearly 47 out of every 100 migrants living outside their countries of birth. Since then, the female proportion of international migration has risen slightly, to reach 48 per cent in

1990 and nearly 49 per cent in 2000 (International Labour Organization 2003). While there has been no major change in the percentage of women and men moving internationally overall, there have been changes in patterns of migration – with more women migrating independently and as main income-earners instead of following male relatives (Martin 2005). There have also been changes in patterns between different regions and countries. In internal sphere generally there are four types of migration that occur namely, rural to rural, rural to urban, urban to urban, urban to rural. In case of developing countries India, people are generally migrating rural to urban. It accounts for roughly 62 per cent of all movements in India 1999–2000 (Deshingkar, 2005). In developing countries, women tend to move shorter distances than men. According to the latest information of NSS 55th round, the estimated number of migrants in India during 1999-2000 was 245.1 million, of which about 77 per cent were females and 23 per cent males. The present paper discusses the gender differential of migration and the volume of net migration and net migration rates for both males and females in major states of India in 2001.

Literature Reviews

Migration is a third demographic component, which plays an important role in changing country's population after fertility and mortality. It influences the age and sex composition of population. Zachariah and Irudaya Rajan (2001) studied gender differentials in migration. They analyzed how women differ from men in the migration process, in terms of levels, trends, characteristics and impacts. Piper Nicola (2001) discussed main issues on gendered migration in the context of eight broad themes set out by the Global Commission on International Migration (GCIM): 1. migrants in the global labour market; 2. migration, development and poverty reduction; 3. irregular migration, state security and human security; 4. migration, development and poverty reduction; 5. migrants in society; migration and human rights; 6. migration and health; 7. the legal and normative framework of international migration; 8. governance of migration. Crush and Williams (2001) also studied gender differential in migration and discrimination. Curran 2003, studied rural-urban migration in the Thai context, net of other factors, and that there are important gender dimensions related to those effects. Chand (2005) analysed the trends in internal migration in India based on NSS data. He discussed the various aspects of internal migration such as migration streams, inter-state flow of migration, employment status of migrants, reasons of migration etc. Kulkarni (1985) while analyzing the census data finds that there is considerable internal migration in India, over a third of the population has moved at least once and over a tenth has moved during a decade. In terms of volume, most of the migrants are females and these are mostly due to marriage. Singh and Mukherji (1986) estimated intercensal migration for Maharashtra state using the place of birth survival technique during the period 1961-71. He estimated migration pattern by age and sex. Gupta (2004) attempted to estimate the volume of net migration and migration rates in northern states of India. She used census survival ratio method for estimating net migration and migration rates.

Objective

The major objectives of this study are:

- To study the gender differential in migration in India during the period 1991 and 2001;
- To study the flow of migration among the major states of India;
- To estimate intercensal net migration and net migration rates during the period of 1991 to 2001.

Data Source

This study is primarily based on the information gathered from secondary sources. These are:

- Age Distribution of Census, 1991 and 2001.
- Migration table (D series) for collecting in migration and out migration according to place of last residence 0 to 9 years).
- Sample Registration System (SRS).

Methodology

Methodology broadly divided into following categories. These are:

- Estimating volume of net migration through Vital Statistics Method.
- Estimating migration rates for selected major states in India.

Net intercensal migration= $(P_{t+n} - P_t) - (B\text{-}D)$

P_{t+n}, P_t = the total population at two successive censuses of the area.

B and D are the number of births and deaths occurred to the residents of the area during the intercensal period. (For this purposes CBR and CDR have been taken from SRS).

$$\textbf{Net migration rate} = \frac{\text{Volume of net migration}}{\text{Mid Year Population}} * 1000$$

Findings and Discussions

Gender Differential in Migration

A person is treated as migrant if he/she had stayed continuously for six months or more at a place (village/town) other than the village/town where he/she was enumerated. People are generally migrated from rural to rural, rural to urban, urban to urban and urban to rural. In Table 8.1 only 4.4 million out of 16.8 million migrants coming from out side the state belong to this stream of rural to rural migration. It is observed that nearly 38 per cent people migrate from rural to urban stream followed by urban to urban stream (27%), rural to rural (26%) and urban to rural (6%). The reasons of migration observed in Table 8.2, indicates that 65 per cent female migrated for marriage purposes, whereas 38 per cent of male migrated for employment purposes. It indicates that work or employment was the most important reason for male migrant and marriage for female migrant. The other important reasons of migration are moving with household and movement after birth.

Tables 8.3 and 8.4 gives information about the age-sex distribution of the inter state inmigrants in India in 1991 and 2001. In this table it is observed that more than half of the migrants are female in 1991 and 2001 (67 per cent of females in comparison with 33 per cent are males). In case of female migration it can be observed that in the age group of 20-24 and 25-29 the proportion of female migration is increasing. It may be because of marriage reasons. Because in this age mostly women will get married and migrate to the husband's place. There may be some women who are migrating for the reason of employment, education or others. In case of male migration age group 20-25 and 40-59 is showing quiet high proportion of male migrants. It may be because these age groups are more suitable for the work or employment and may be they are migrating for that purpose. In the age group 0-14 migrants are higher. In the age group of 60+ the proportion of migrants is reduced for both the sex. This type of pattern is observed in both 1991 and 2001.

Interstate flow of migrants in major states is observed in Tables 8.5, 8.6 and 8.7 respectively. Most of the states received the bulk of interstate migrants from the neighbouring states. For instance, the largest proportion of male and female migrants into Andhra Pradesh are from Karnataka (29.44% female) and (28/90% male) migrants. Similarly largest proportion of migrants into Karnataka is from Maharashtra, Andhra Pradesh and Tamil Nadu. The Maharashtra state is receiving largest proportion of migrants from Gujurat (62.59% male and 73.22% female). The largest proportion of migrants of Haryana is from Punjab and vice versa.

The volume of net migration and migration rates are estimated for the major states of India in the Table 8.8. Among the major states the number of estimated migrants was the highest in Uttar Pradesh, followed by Maharashtra, Madhya Pradesh, Andhra Pradesh and West Bengal. The states where the number of estimated migrants was the lowest were Haryana, Punjab and Assam. However, the size of the population in different states has a bearing on the magnitude of migration. The net migration rate has been estimated by the number of persons of that category in that region during the specified period per 1000 person of that category in that region. Net migration rate is highest in Uttar Pradesh (624.56 migrants per 1000 person) followed by Haryana, and Rajasthan (569.76 and 569.61 migrants per 1000 persons). The lowest net migration rates have been seen in Andhra Pradesh and Bihar (238.74 and 333.88) respectively.

Table 8.1: Inter state migration stream (Duration 0-9), India 2001

Migration Stream	Persons	Males	Females	2001 (in Per cent)		
				Persons	Males	Females
Total	16,826,879	8,512,161	8,314,718	100.00	100.00	100.00
Rural-Rural	4,474,302	1,759,523	2,714,779	26.6	20.7	32.7
Rural-Urban	6.372,955	3,803,737	2,569,218	37.9	44.7	30.9
Urban-Rural	1,053,352	522,916	530,436	6.3	6.1	6.4
Urban-Urban	4,490,480	2,201,882	2,288,598	26.7	25.9	27.5
Unclassified	435,790	224,103	211,687	2.6	2.6	2.5

Source: Table D-2, Census of India 2001.

Table 8.2: Percentage distribution of reasons for migration by last residence with duration (0-9years) India 2001

Reason for migration	*Persons*	*Males*	*Females*
Work/Employment	14.7	37.6	3.2
Business	1.2	2.9	0.3
Education	3.0	6.2	1.3
Marriage	43.8	2.1	64.9
Moved after birth	6.7	10.4	4.8
Moved with household	21.0	25.1	18.9
Other	9.7	15.7	6.7

Source: Census of India 2001.

Table 8.3: Gender Differential in migration in 1991

Age-Group	*Males*	*Females*
All ages	33.47	66.53
0-14	59.17	53.33
15-19	32.78	79.72
20-24	16.98	95.53
25-29	21.95	90.56
30-34	45.21	67.29
35-39	59.37	53.12
40-59	66.62	45.87
60+	51.62	60.88

Source: Table D-2, Census of India 1991.

Table 8.4: Gender differential in migration in 2001

Age Group	*Males*	*Females*
All ages	33	67
0-14	58.65	53.65
15-19	61.25	51.02
20-24	16.54	96.46
25-29	21.42	91.50
30-34	44.41	68.13
35-39	58.46	53.85
40-59	65.41	46.79
60+	49.43	63.03

Source: Table D-2, Census of India 2001.

Table 8.5: Inter state flow of female migrants among major states—2001

States	*A.P*	*Assam*	*Bihar*	*Gujarat*	*Haryana*	*Karnataka*	*Kerala*	*M.P*	*Maharashtra*	*Orissa*	*Punjab*	*Rajasthan*	*T.N*	*U.P.*	*W.B.*	*Others*
Andhra Pradesh		0.32	1.61	0.99	0.31	21.36	3.03	1.06	11.45	7.90	0.39	1.94	13.15	1.59	1.79	2.96
Assam	0.51		7.70	0.27	0.50	0.24	0.45	0.29	0.44	0.43	0.64	1.65	0.29	2.89	8.71	12.23
Bihar	0.54	3.50		1.76	0.61	0.25	0.10	0.18	1.08	1.29	0.34	0.47	0.08	11.22	3.58	18.99
Gujarat	3.63	0.96	9.85		1.72	1.77	3.65	13.76	62.59	5.80	1.44	31.97	2.06	31.57	3.44	6.53
Haryana	0.62	0.75	12.56	0.63		0.33	0.59	2.11	1.39	0.65	26.85	33.67	0.47	49.18	3.64	38.11
Karnataka	29.44	0.35	1.07	1.34	0.64		13.13	0.58	22.83	0.62	0.46	3.09	23.04	1.91	1.44	3.61
Kerala	1.93	0.20	0.38	0.89	0.24	8.04		0.62	4.24	0.43	0.26	0.53	28.46	0.78	0.57	2.79
Madhya Pradesh	2.32	0.30	3.73	2.57	1.05	0.58	1.23		18.00	0.82	1.11	16.30	0.54	37.02	1.65	13.81
Maharastra	19.83	0.89	10.17	26.57	1.72	50.04	8.91	28.86		3.25	2.31	14.03	8.48	56.78	7.47	23.93
Orissa	11.72	0.38	6.53	0.74	0.34	0.30	0.73	1.20	1.03		0.48	0.82	0.60	2.32	12.59	28.52
Punjab	0.72	1.05	11.54	0.76	27.79	0.50	0.59	1.77	1.98	0.69		10.98	0.59	29.25	3.24	32.47
Rajasthan	0.87	0.53	3.07	6.98	17.71	0.55	0.77	18.14	2.03	0.41	6.61		0.53	16.11	2.11	5.11
Tamil Nadu	10.05	0.20	0.61	0.63	0.14	9.85	13.80	0.49	3.42	0.38	0.20	1.43		0.80	0.73	1.94
Uttar Pradesh	0.18	1.46	7.82	0.34	2.35	0.23	0.30	7.02	0.73	0.38	1.01	2.66	0.15		2.06	14.56
West Bengal	1.22	8.05	37.85	0.85	0.66	0.44	0.61	0.91	1.52	7.18	1.00	2.02	0.62	8.89		33.59

Source: Table D-2, Census of India 2001.

Table 8.6: Inter state Flow of Male Migrants among Major States—2001

States	*A.P*	*Assam*	*Bihar*	*Gujarat*	*Haryana*	*Karnataka*	*Kerala*	*M.P*	*Maharashtra*	*Orissa*	*Punjab*	*Rajasthan*	*T.N*	*U.P.*	*W.B.*	*Others*
Andhra Pradesh	0	0.45	3.26	1.41	0.46	10.23	2.94	1.38	9.51	8.55	0.53	3.52	11.48	3.01	2.92	4.08
Assam	0.66	0.00	18.17	0.29	0.54	0.23	0.49	0.35	0.46	0.88	0.80	2.50	0.38	3.69	8.01	13.44
Bihar	0.06	0.31	0.00	0.21	0.13	0.04	0.03	0.06	0.22	0.18	0.29	0.11	0.03	1.18	0.85	3.11
Gujarat	4.98	1.53	41.29	0.00	2.48	2.44	5.18	20.70	73.22	34.21	2.11	52.33	3.36	89.78	8.18	14.58
Haryana	0.97	1.20	52.05	1.07	0.00	0.56	0.97	4.22	2.50	2.35	23.35	27.56	0.87	98.54	11.01	53.23
Karnataka	28.90	0.74	4.54	2.11	0.86	0.00	16.98	1.13	17.29	2.94	0.79	6.18	29.89	4.91	4.28	6.38
Kerala	2.07	0.32	0.70	1.01	0.24	7.19	0.00	0.63	4.58	1.02	0.27	0.80	35.47	1.11	1.06	2.78
Madhya Pradesh	1.85	0.42	7.53	2.38	1.29	0.64	1.30	0.00	11.49	1.51	1.45	11.37	0.82	29.32	2.82	16.90
Maharastra	25.06	1.84	46.81	29.96	2.59	59.25	15.19	34.86	0.00	10.87	3.56	30.71	13.80	168.03	24.19	38.59
Orissa	5.62	0.33	6.56	1.70	0.31	0.29	0.59	0.74	0.95	0.00	0.42	0.68	0.63	2.22	8.63	11.53
Punjab	0.79	1.30	56.44	0.90	15.44	0.90	0.76	2.57	2.59	1.93	0.00	9.36	0.76	76.06	7.77	39.58
Rajasthan	0.86	0.62	7.62	5.10	6.03	0.55	0.69	9.00	3.39	0.78	4.56	0.00	0.65	13.85	3.78	5.58
Tamil Nadu	7.80	0.25	1.31	0.63	0.16	5.82	11.02	0.59	3.23	0.96	0.22	1.80	0.00	1.08	1.21	2.03
Uttar Pradesh	0.12	0.48	4.48	0.21	0.60	0.11	0.17	1.22	0.45	0.24	0.65	0.64	0.11	0.00	0.93	8.20
West Bengal	0.99	5.40	43.96	0.81	0.60	0.37	0.47	0.70	1.42	6.61	0.86	1.76	0.55	9.66	0	20.37

Source: Table D-2, Census of India 2001.

Table 8.7: Inter State Flow of Migrants among Major States—2001

States	*A.P*	*Assam*	*Bihar*	*Gujarat*	*Haryana*	*Karnataka*	*Kerala*	*M.P*	*Maharashtra*	*Orissa*	*Punjab*	*Rajasthan*	*T.N*	*U.P.*	*W.B.*	*Others*
Andhra Pradesh		0.21	1.50	0.65	0.21	4.69	1.35	0.63	4.36	3.92	0.24	1.61	5.26	1.38	1.34	1.87
Assam	0.30		8.14	0.13	0.24	0.10	0.22	0.16	0.21	0.39	0.36	1.12	0.17	1.65	3.59	6.02
Bihar	0.04	0.19		0.13	0.08	0.02	0.02	0.04	0.14	0.11	0.18	0.07	0.02	0.74	0.53	1.95
Gujarat	2.23	0.68	18.51		1.11	1.10	2.32	9.28	32.83	15.34	0.95	23.46	1.51	40.25	3.67	6.54
Haryana	0.34	0.42	18.18	0.37		0.20	0.34	1.47	0.87	0.82	8.16	9.63	0.30	34.42	3.84	18.59
Karnataka	12.96	0.33	2.04	0.94	0.39		7.61	0.51	7.75	1.32	0.35	2.77	13.41	2.20	1.92	2.86
Kerala	1.03	0.16	0.35	0.50	0.12	3.59		0.31	2.29	0.51	0.13	0.40	17.72	0.56	0.53	1.39
Madhya Pradesh	0.76	0.17	3.09	0.98	0.53	0.26	0.53		4.72	0.62	0.60	4.68	0.34	12.05	1.16	6.95
Maharastra	10.86	0.80	20.30	12.99	1.12	25.69	6.59	15.11		4.71	1.54	13.31	5.98	72.85	10.49	16.73
Orissa	3.25	0.19	3.79	0.98	0.18	0.17	0.34	0.43	0.55		0.24	0.40	0.36	1.29	4.99	6.66
Punjab	0.32	0.53	22.98	0.37	6.28	0.37	0.31	1.05	1.05	0.79		3.81	0.31	30.96	3.16	16.11
Rajasthan	0.40	0.29	3.55	2.37	2.81	0.26	0.32	4.19	1.58	0.36	2.12		0.30	6.45	1.76	2.60
Tamil Nadu	4.04	0.13	0.68	0.33	0.08	3.01	5.70	0.30	1.67	0.50	0.11	0.93		0.56	0.63	1.05
Uttar Pradesh	0.07	0.27	2.53	0.12	0.34	0.06	0.09	0.69	0.26	0.13	0.37	0.36	0.06		0.52	4.63
West Bengal	0.52	2.85	23.21	0.43	0.32	0.19	0.25	0.37	0.75	3.49	0.45	0.93	0.29	5.10		10.76

Source: Table D-2, Census of India 2001.

Table 8.8:	Mid Year Population			Net Migration			Net Migration Rates		
States				*Male*	*Female*	*Total*	*Male*	*Female*	*Total*
Andhra Pradesh	72251994	70466021	142718015	17161110	16912561	34073671	237.5175	240.0102	238.7482
Assam	12717513	11817412	24534925	17779222	19840823	12454871	1398.011	1678.948	507.6384
Bihar	44222943	40463544	84686487	14605883	13669840	28275723	330.2784	337.831	333.8871
Gujarat	23870393	22119907	45990300	12855924	11643291	24499215	538.572	526.3716	532.704
Haryana	10095714	8708393	18804106	5771282	4942691	10713973	571.6567	567.5779	569.7678
Karnataka	24925418	23988464	48913882	12357666	11997463	24355130	495.7857	500.1347	497.9186
Kerala	14878805	15591142	30469946	6415781	6990145	13405926	431.2027	448.3408	439.9721
Madhya Pradesh	38721142	35611007	74332149	21464850	19089010	40553860	554.3445	536.0424	545.5763
Maharastra	45609007	42290874	87899881	24540606	22337314	46877920	538.0649	528.1828	533.3104
Orissa	17362358	16869840	34232198	8483085	8263460	16746545	488.5906	489.8363	489.2045
Punjab	11881540	10438945	22320484	6156591	5352709	11509300	518.1644	512.7635	515.6385
Rajasthan	26231396	24025194	50256589	14821193	13805876	28627069	565.0173	574.6416	569.6182
Tamil Nadu	29849942	29282371	59132313	13472014	13544073	27016086	451.3246	462.5333	456.8752
Uttar Pradesh	75355285	67419357	124987150	40659046	37403893	78062940	539.5646	554.7946	624.5677
West Bengal	38488309	35638772	74127081	18968123	18077087	37045211	492.8282	507.2309	499.7527

Source: Table D-2, Census of India 2001.

Graph 8.1: Gender Dimension in Migration–1991

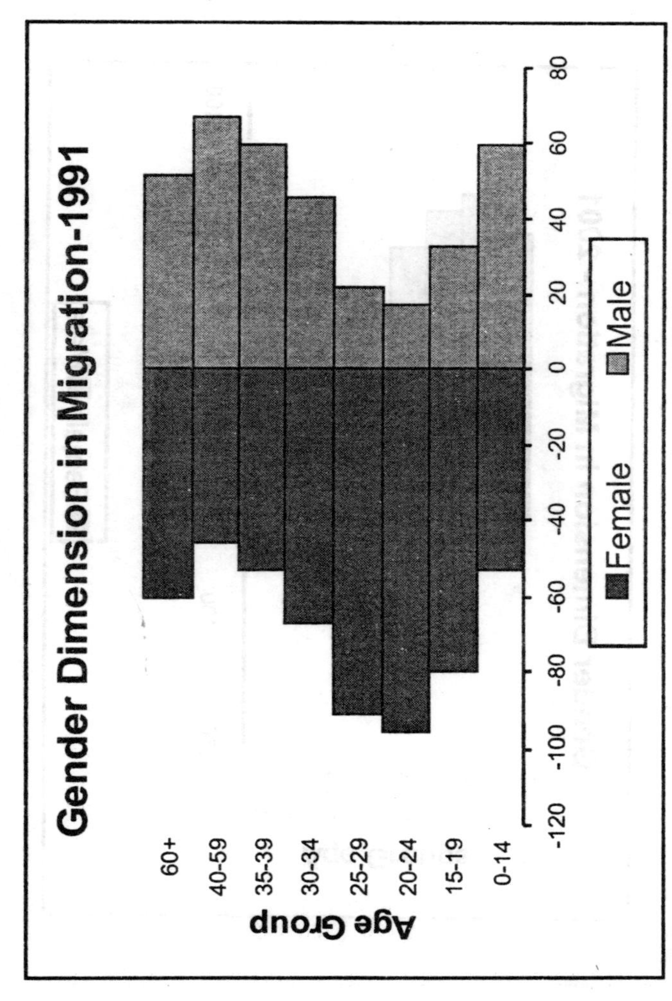

Source: Table D-2, Census of India 2001.

Graph 8.2: Gender Dimension in Migration 2001

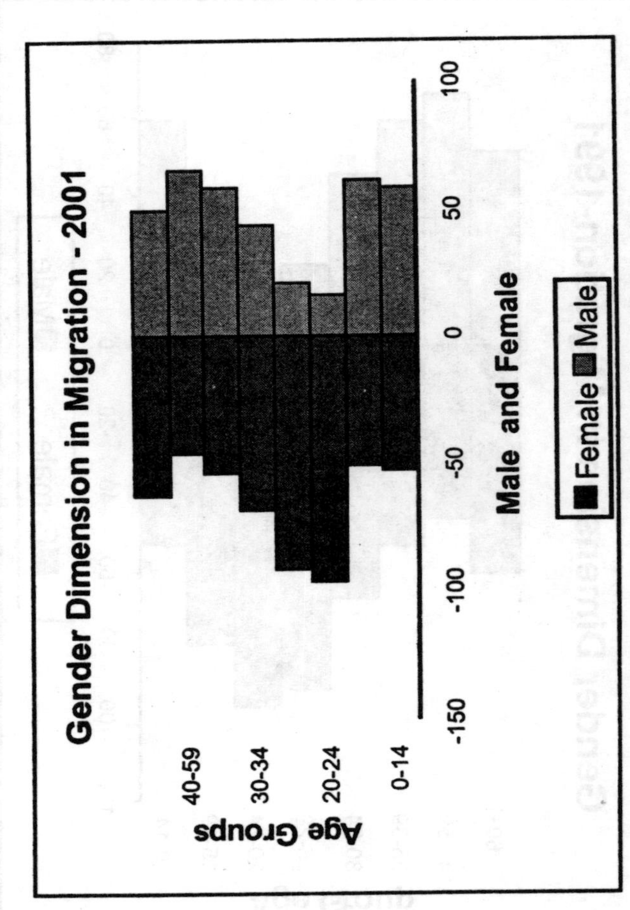

Source: Table D-2, Census of India 2001.

Conclusion

Gender plays an important role in internal and international migration. Most decisions to migrate are made in response to a combination of economic, social and political pressures and incentives. Inequalities within and between countries create incentives to move. The push and pull factors play an important role for migration in case of India. These factors are marriage, higher employment opportunities, education proposes, better standard of living etc. It is seen that in India, most of the migration takes place due to marriage and short distance only. People migrated rural to rural and rural to urban streams. Due to lack of unemployment in the place of origin, people are migrated higher to urbanized states. It is observed that, Maharashtra, Gujurat, Karnataka, Andhra Pradesh and Tamil Nadu have been receiving more inmigrants in comparison with other states. The inter-state migration pattern reflects that there is an inequality in the regional development. Some states experience high inmigration. At the same time, the backward states like U.P, Bihar, M.P, etc are experiencing heavy out-migration. At the policy point of view, there should be a balanced migration in the context of development.

REFERENCES

Census of India 1991, Social Cultural Table, Registrar General and Census Commissioner, India

Census of India 2001, Social Cultural Table. Registrar General and Census Commissioner, India.

Chand Himal (2005), "Migration in India—An Overview of Recent Evidence", *Man and Development, September 2005.*

Crush, J. and Williams, V. (eds), 2002, "Policing Migration: Immigration Enforcement and Human.

Rights in South Africa", *Migration Policy Brief* No. 14, Southern African Migration Project.

Deshingkar, P., 2005, "Maximising the Benefits of Internal Migration for Development", Background.

Paper Prepared for the Regional Conference on Migration and Development in Asia, Lanzhou, China, 14–16 March 2005.

Gupta, K. (1993); "Women Migrants and their Economic Roles in Indian Metropolises", *Dynamics of Population and Family Welfare, pp. 161-186.*

International Migration Report 2002, United Nations, New York, 2002.

Kulkurni, P.M. (1985); "International Migration and Urbanization in India—Trends and Policy Interventions," *IIPS, National Seminar.*

Martin, S., 2005, "World Survey on the Role of Women in Development: Women and International Migration", *New York: United Nations Department of Economic and Social Affairs and Division for the Advancement of Women.*

Piper, N., 2005, "Gender and Migration", Background *Paper for Global Commission on International.*

Migration (GCIM) and Appendix to the GCIM Global Report on Migration.

Singh D.P.and Mukherji S.(1986), " Estimation of Intercensal Migration from Place of Birth Data: A Study of Maharashtra State 1961-71", *Paper Presented in National Seminar on Migration Research in the Context of Development, sponsored by the Universal Grants Commission*

Sample Registration System (SRS), *Registrar General of India, New Delhi.*

World Migration Report 2003, " Managing Migration Challenges and Responses for People on the Move", *Vol. 2, IOM, World Migration Report Series 2003.*

Zachariah and Irudaya Rajan (2001), "Gender Dimension of Migration in Kerala: Macro and Micro Evidence", *Asia Pacific Population Journal, Vol. 16, No. 3, September, 2001.*

9

Gender Inequality in the Length of Life in Major States of India

Mr. Sangram Kishor Patel and
Dr. L. Ladusingh

Introduction

In addressing the question "what would we like to be equally distributed in the population", several ethical and technical issues arise (Gakidou, et. al., 2000). Would we consider that perfect equality exits when all individuals live the same number of years? When they enjoy the same level of health? When they have the same health status at all points in their lives? Considering these questions, Murray et. al. (1999) have defined health inequality as the variation in health status across individuals in a population. Average achievement is no longer considered a sufficient indicator of a country's performance on health; rather, the distribution of health in the population is also key and should be measured as a distinct dimension of the performance of health systems (Murray, et. al., 2000).

Inequalities in health, both between and within population are a major public concern in developing countries. Knowledge of inequality of health among poor, non-poor, rural-urban, ethnic groups, sex, age, geographical regions etc. are essentials for planning and implementation of health programmes on equity basis. In the period when centralized planning was accepted as a key instrument of development in India, one of the basic objectives stipulated in National Health Policy (NHP-2002) was to reduce inequality in health services among poor, non-poor and promote quality of life. Health inequalities are prominent in the policy agenda (Acheson, 1998; Barker, 1990; Kaplan and Lynch, 1997; Kunst, et. al. 1995; Murray, et. al. 1999; Wagstaff, 1991; Mormot, 1991).

Despite widespread reorganization of this magnitude in many high and low income countries, there is considerable debate about the measuring and measurement of health inequalities and social group differences. The lack of standard definition, measurement strategies and indicators have limited and will continue to limit comparisons between and within countries and over the time – of health inequalities and perhaps more importantly, comparative analyses of their determinants. Comparative studies are essential for formulating effective policies, which would enable governments to reduce these inequalities (Murray, et. al, 1999). We need better methods for measuring health equity in order to place health inequalities at the center of the policy debate. This seems to be the only way to ascertain the true magnitude of the problems and of maintaining programme towards it's solution (Gakidou, et. al. 2000).

According to the WHO (1948), health is a state of complete physical, mental and social well being and not merely the absence of disease or infirmity. The health status of a population is a significant indicator of human development. Good health affects several aspects of life and personal well being. A healthy population will have high productivity and thereby has a bearing on a country's economy. Among the different basic health aspects of population mortality is an important phenomenon. Indicators such as infant mortality rate, life expectancy etc. are used to measure health status. Mortality is mostly used to evaluate the health status of a population, since it is comparatively simple to analyze and data are easily available.

Death is the great leveler and operator at different levels in different places and among different groups. Social class influences mortality levels. This inequality of death is part of the gross inequality in health status between countries and within countries and equally unacceptable (WHO, 1980). There is a long tradition of the use of the length of life for expressing social inequalities in mortality. Antonovsky (1967) viewed many of the studies on socio-economic gradients in length of life, which were completed beginning from the 19th century. However, in the second half of the 20th century this tradition has weakened. Antonovsky quotes "Very few later investigators have dealt with class difference in life expectancy, preferring to concentrate on differences in mortality rates" (Antonovsky, 1967).

Certainly, simplicity of mortality rates and of their linear aggregates (e.g. standardized mortality rates and ratios) as well as their dominance in epidemiology (relative risks and odd ratios) are important advantages. However, such indicators reflect only frequency (or intensity) of death and do not directly indicate its prematureness (Shkolnikov, et al. 2001). Life expectancy directly reflects prematureness of death, revealing a prevalence of younger or older ages at death in different social groups.

A number of summary indices of inequality measure the general level of inter-group inequality in a given population. However, a number of such indices have been constructed with respect to mortality or incidence rates by Kunst et. al, (1994). Similar indices have been developed by Shkolnikov et. al (2001) for life expectancies. But these procedures have not been widely used so far.

Need for the Study

For any country the inequalities in health of the population is an important issue. Both from the point of view of the individuals and the nation, there is a necessity for reducing the health inequalities among different population groups. Knowledge of geographical areas, population groups which contribute most to the inequalities in population health be potential inputs for policies and programmes. Improvement of the health status and reducing the health inequalities among different population groups is a priority of the government's plans and programmes for most of the countries. Further, comparative studies across populations essential for formulating effective policies with which governments will able to take better action and reduce health inequalities (Murray, et. al, 1999).

In India various researchers have analyzed, decomposed life expectancy to understand the trends, patterns and population projection purpose (to ascertain the mortality situation). But studies based on population inter group inequalities in the length of life are scarce and have not been explored in India. It is always better to measure the inequalities among different population subgroups in smaller regions (states and districts level). Further it is also important to capture the real magnitude of inequalities among these population subgroups in smaller regions. Because small-area analyses hold out the promise of being one of the most refined methods for revealing the underlying distribution of the life expectancy in the population groups. In this study we make an attempt to explore some of these issues so that the study can produce better output for the policy implication.

Objective

The main objective of the study is to measure the gender (Male/Female) inequality in the length of life in major states of India.

Methodology

Shkolnikov et. al.(2001) have examined inter-group inequalities in the mortality schedule in terms of life expectancy by formulating an approximate linear decomposition of the life expectancy by overall population by population group that permits one to build aggregate indices of inequality in length of life similar to aggregate indices of inequality in mortality rates. The survival function l_x of the life table can be defined as exp $(-\int_0^x \mu_t dt)$. Where μ_t is the over all force of mortality.

Further, the link between life expectancy at age x for the whole population and group-specific mortality schedules can be written as:

$$e_x = \frac{1}{l_x}\int_x^\infty l_t dt = \frac{1}{l_x}\int_x^\infty \left[\exp(-\int_0^t \sum_i p_y^i \mu_y^i dy) \right] dt \qquad \text{----------(1)}$$

Again, the relationship between equation (1) and the life expectancies in specific groups can be written as:

$$e_x^i = \frac{1}{l_x^i}\int_x^{\infty}\left[\exp(-\int_0^t \mu_y dy)\right]dt \qquad \text{------------------------ (2)}$$

Simpler linear decomposition for the overall life expectancy could be achieved through a division of the overall life table cohort into fractions corresponding to specific population groups.

Properties of these fractions are as follows:

1. At every age exact age x the sum of the fractions should be equal to the number of survivors in the overall cohort.
2. The sum of person-years lived by all fractions after age x should be equal to total number of person-years lived by the whole cohort.

We can get e_x^i from the group-specific life tables. Only we have to estimate the values of l_x^i or θ_x^i from following equations,

$$l_x = \sum_{i=1}^{N} l_x^i \qquad \text{------------------------ (3)}$$

$$e_x l_x = \sum_{i=1}^{N} l_x^i e_x^i \qquad \text{------------------------ (4)}$$

Shkolnikov et. al (2001) have formulated two summary indices for measuring the inequality in the life expectancy using the calculated values of θ_x^i.

(i) Population Attributable Life Loss (PALL)

Population attributable life loss (PALL) index in relative terms as a proportion of increase in overall life expectancy that would occur if all groups had the life expectancy of the best group and in absolute terms it is the increase in the overall life expectancy that would be achieved if all groups had the life expectancy of the best group.

$$PALL = \sum_{i=1}^{N} \frac{(e_x^{highest} - e_x^i)\theta_x^i}{e_x}$$

Where,

$e_x^{highest}$ = Life expectancy of the best group at exact age x.

e_x^i = Life expectancy of the ith group at exact age x.

e_x = Overall life expectancy at exact age x.

θ_x^i = Weight of ith group population at exact age x.

(ii) Index of Dissimilarity in Length of Life (IDLL)

The index of dissimilarity in length of life (IDLL) shows the proportion of person-years of the length of life that should be redistributed among the groups if all groups were to achieve the average level of life expectancy. Again, in absolute terms it shows how many years of length of life should be redistributed in the overall population to reach the situation of complete equality.

$$IDLL = \frac{\sum_{i=1}^{N} \left| e_x - e_x^i \right| \theta_x^i}{e_x}$$

where,

e_x= Overall life expectancy of the population at exact age x.

e_x^i = Life expectancy of the ith group population at exact age x.

θ_x^i = Weight of the ith group population at age x.

Data Sources

SRS based Abridged life tables (1998-2002)- 2005: In the absence of dependable data from the Civil Registration System (CRS) in India, the sample Registration System (SRS) has been the main source of information on fertility and mortality indicators at the state and national level. The Office of the Registrar General of India initiated the scheme of sample registration of births and deaths in India popularly known as Sample Registration System (SRS) in 1964-65 on a pilot basis and on full scale from 1969-70. The SRS mechanism involves collection of data through two different procedures viz., continuous enumeration and retrospective half-yearly surveys followed by a process of matching of the two records and subsequent field verification of unmatched and partially matched events. The methodology provides a crosscheck on the correctness and completeness of events of births and deaths listed in both the records. The SRS's main objective is to provide reliable estimates of fertility and mortality indicators at state and national levels for rural and urban areas separately in India. The estimated age-specific death rates derived from the SRS provide the necessary database for undertaking construction of abridged life tables. To adjust the sampling fluctuation and for augmenting the sample size, five-year average has been compiled for rural and urban areas both for male and female population. For the present study, we have used SRS based Abridged life tables (1998-2002)-2005 data, which has been constructed for all India and sixteen major states using the death data of the period 1998-2002. SRS based Abridged life tables have been constructed according to population groups such as sex and place of residence.

Findings

In this study, we measure the magnitude of the inter group gender (male and female) inequalities in the length of life at birth among major states of India in general as well as in absolute terms.

Table 9.1 illustrates the inter group (male and female) inequalities in the length of life in India's total population by different states. According to population attributable life loss (PALL) measure among the states, Kerala has the highest inequalities of length of life of 2.4 years, that could be achieved by the overall population if Kerala males experience mortality conditions characteristic of the Kerala females. In other words, it is 3.3 per cent of the overall number of years lived by the overall

Table 9.1: Population attributable life loss and Index of dissimilarity in length of life at birth by sex in major states of India, 1998-2002

	Total	*Male*	*Female*				
States	e_x	e_x	e_x	*PALL(%)*	$PALL_{abs}$	*IDLL(%)*	$IDLL_{abs}$
Andhra Pradesh	63.5	62.0	64.6	1.7	1.1	2.0	1.3
Assam	57.9	57.7	58.1	0.3	0.2	0.3	0.2
Bihar	60.8	61.4	59.5	2.1	1.3	1.4	0.8
Gujarat	63.4	62.4	64.4	1.6	1.0	1.6	1.0
Haryana	65.2	64.7	65.4	0.3	0.2	0.4	0.3
Himachal Pradesh	65.9	65.7	66.3	0.6	0.4	0.4	0.3
Karnataka	64.5	62.8	66.2	2.6	1.7	2.6	1.7
Kerala	73.5	70.8	75.9	3.3	2.4	3.5	2.5
Madhya Pradesh	56.9	57.0	55.4	2.6	1.5	0.3	0.2
Maharashtra	66.2	65.0	67.4	1.8	1.2	1.8	1.2
Orissa	58.5	58.4	58.5	0.0	0.0	0.0	0.0
Punjab	68.5	67.4	69.5	1.5	1.0	1.5	1.0
Rajasthan	61.1	60.5	61.6	0.8	0.5	0.9	0.5
Tamil Nadu	65.2	64.2	66.3	1.7	1.1	1.6	1.0
Uttar Pradesh	59.1	59.4	58.5	1.0	0.6	0.7	0.4
West Bengal	63.9	63.3	64.8	1.4	0.9	1.1	0.7

cohort. After Kerala, Karnataka and Madhya Pradesh follow with inequalities in the length of years of 1.7 years (2.6 per cent overall number of years) and 1.5 years (2.6 per cent), respectively. At the same time, results show that Orissa males and females do not have any inequality in the length of life.

According to Index of dissimilarity in length of life (IDLL) measure, again Kerala is in the top with 2.5 years of length of life (or 3.5 per cent of the overall life expectancy) should be redistributed to achieve the situation of complete equity between male and female groups at the level of life expectancy of the overall population of Kerala. Karnataka and Andhra Pradesh succeedingly coming thereafter with 1.7 years (2.6 per cent) and 1.3 years (2 per cent) respectively. As per PALL, IDLL measure also shows that Orissa does not have any inequality in the length of life between males and females. Assam, Madhya Pradesh, Haryana and Himachal Pradesh show considerably less number of years of inequalities in the length of life, which vary from 0.2 to 0.3 years of length of life with 0.3 per cent to 0.4 per cent of over all life expectancy, respectively.

Table 9.2 depicts the intergroup (male and female) inequalities in the length of life in rural India population. This table does not show any significant difference from the Table-9.1 result. The illustration of the result of Table-9.2 is more or less similar to the Table-9.1.

Table 9.3 shows the intergroup (male and female) inequalities in the length of life in urban India population. According to PALL measure, both Kerala and Tamil Nadu possess the highest inequalities with 2.1 years of length of life each and which could be achieved by the overall population if both states males have to go through the mortality conditions of females in each state. But when it comes in the account of percentage of the overall number of years lived by the overall birth cohort, Tamil Nadu is highest with 3 per cent followed by Kerala with 2.8 per cent. In absolute terms Rajasthan shows lowest inequality of 0.3 years of length of life, which account 0.5 per cent of over all life expectancy.

Table 9.2: Population attributable life loss and Index of dissimilarity in length of life at birth by sex in major states of rural India, 1998-2002

	Total	Male	Female				
States	e_x	e_x	e_x	PALL(%)	$PALL_{abs}$	IDLL(%)	$IDLL_{abs}$
Andhra Pradesh	62.4	61.0	63.4	1.6	1.0	1.9	1.2
Assam	57.1	56.9	57.3	0.4	0.2	0.4	0.2
Bihar	59.9	60.9	58.8	1.8	1.1	1.7	1.0
Gujarat	62.1	61.2	62.9	1.3	0.8	1.4	0.8
Haryana	64.4	64.4	64.4	0.0	0.0	0.0	0.0
Himachal Pradesh	65.8	65.5	65.9	0.2	0.1	0.2	0.2
Karnataka	62.8	61.1	64.7	3.0	1.9	2.9	1.8
Kerala	73.4	70.9	75.6	3.0	2.2	3.2	2.3
Madhya Pradesh	55.6	55.8	55.4	0.4	0.2	0.4	0.2
Maharashtra	64.4	63.1	65.3	1.4	0.9	1.7	1.1
Orissa	57.8	57.8	57.8	0.0	0.0	0.0	0.0
Punjab	67.7	66.9	68.5	1.2	0.8	1.2	0.8
Rajasthan	59.7	59.5	59.8	0.2	0.1	0.2	0.1
Tamil Nadu	63.8	62.7	64.6	1.3	0.8	1.5	0.9
Uttar Pradesh	58.2	58.9	57.6	1.0	0.6	1.1	0.6
West Bengal	62.5	61.7	63.4	1.4	0.9	1.4	0.8

Table 9.3: Population attributable life loss and Index of dissimilarity in length of life at birth by sex in major states of urban India, 1998-2002

States	Total	Male	Female				
	e_x	e_x	e_x	PALL(%)	$PALL_{abs}$	IDLL(%)	$IDLL_{abs}$
Andhra Pradesh	67.0	65.2	68.5	2.2	1.5	2.4	1.6
Assam	66.6	66.1	67.3	1.1	0.7	0.9	0.6
Bihar	66.8	66.1	67.5	1.0	0.7	1.0	0.7
Gujarat	65.6	64.5	67.5	2.9	1.9	2.1	1.4
Haryana	68.6	66.7	70.2	2.3	1.6	2.5	1.7
Himachal Pradesh	68.3	66.7	69.9	2.3	1.6	2.3	1.6
Karnataka	68.9	67.6	69.8	1.3	0.9	1.5	1.1
Kerala	74.1	70.2	76.2	2.8	2.1	3.7	2.7
Madhya Pradesh	64.3	63.3	65.1	1.2	0.8	1.4	0.9
Maharashtra	70.3	68.7	72.0	2.4	1.7	2.3	1.6
Orissa	65.6	63.5	66.9	2.0	1.3	2.4	1.6
Punjab	71.0	68.9	72.0	1.4	1.0	1.9	1.4
Rajasthan	66.1	65.9	66.4	0.5	0.3	0.4	0.2
Tamil Nadu	68.9	66.9	71.0	3.0	2.1	3.0	2.0
Uttar Pradesh	63.1	62.4	63.7	1.0	0.6	1.0	0.6
West Bengal	69.0	68.0	70.3	1.9	1.3	1.6	1.1

Further, according to IDLL measure of inequality in the length of life in urban population, Kerala has the highest number of years of inequalities in the length of life with 2.7 years (3.7 per cent of the over all life expectancy), which should be redistributed to achieve the complete equity between males and females at the level of life expectancy of the overall population of Kerala. Tamil Nadu follows Kerala in the inequalities in the length of life with 2 years (3 per cent of the over all life expectancy) and should be redistributed to achieve equity among over all population.

Again Rajasthan shows the lowest inequalities in the length of life among major states urban population in India.

Conclusion

It is clear from the study that the population groups, which have the larger gap between the life expectancies, are those states which have the highest magnitude of inequalities in the length of life. The inter group inequalities in the length of life at birth between male and female, Kerala has the highest inequalities. Although Kerala has the highest life expectancy in the country and, at the same time, it is one of the lowest mortality rate state. But there is largest gap between male and female life expectancy at birth in Kerala. Karnataka, Madhya Pradesh and Andhra Pradesh follow the list among the highest inequalities in the length of life states. The inequalities gap between male and female in the length of life picture is totally different in the context of urban population as a whole in India. Gujarat, Tamil Nadu and Maharashtra come after Kerala in the highest inequalities among the major states. One important finding from the study is that Orissa could not be able to count the inequalities in the length of life between male and female of rural region as well as considering population as a whole. Plausible reason in both the cases is that life expectancy of male and female are nearly same. But the urban figures show a different picture, Orissa accounts for a considerable inequity in the length of life between male and female considering urban population only. Rajasthan possesses the lowest inequality in the length of life between male and female in the urban context. Kerala has widest variation in the inequalities among the major states in India. But as rural and urban group population has nearly the same life expectancy, they have the lowest magnitude of inequality.

Life expectancy directly reflects prematureness of death, revealing a prevalence of younger or older ages at death in different population groups. The evidences obtained from this study suggest that there is need to focus the inequalities in the length of life in India. Males as well as females of rural areas are in more vulnerability of death. The results based on the magnitude of inter group inequalities always have better inputs for the efficient and effective policies and programme management rather than only the difference in the length of life.

REFERENCES

Acheson, D, 1998. *Independent Enquiry into Inequalities in Health*. The Stationary Office. London.

Antonovsky, A, 1967. Social Class, Life Expectancy and Overall Mortality. *Milbank Memorial Fund Quarterly*, 45 (2): 31-73.

Barker, J, 1990. Inequalities—Whose Health for All? *Health Visit*, 63(7): 232-233.

Gakidou, EE, CLJ Murray and J. Frenk, 2000. Defining and Measuring Health Inequality: An Approach Based on the Distribution of Health Expectancy. *Bulletin of the World Health Organisation*, 78 (1): 42-54.

Kaplan, G and J. Lynch, 1997. Whither Studies on the Socio-economic Foundations of Population Health? *American Journal of Public Health*, 87: 1409-1411.

Kunst, A.E, Mackenbach, J.P, 1994. *Measuring Socio-economic Inequalities in Health*, WHO Regional Office for Europe, Copenhagen.

Kunst, AE, 1995. International Variation in Socio-economic Inequalities in Self Reported Health. *Journal of Epidemiology and Community Health*, 49(2): 1487-1491.

Marmot, MG, 1991. Health Inequalities Among British Civil Servants: The Whitehall II Study. *Lancet*, 337: 1387-1393.

Murray, C.J.L, J.A. Salomon and C. Mathers, 2000. A Critical Examination of Summary Measures of Population Health. *Bulletin of the World Health Organisation*, 78 (8): 981-993.

Murray, CLJ, 1998. *U.S. Patterns of Mortality by Country and Race: 1965-1994*. Havard School of Public Health and National Center for Disease Prevention and Health Promotion, Cambridge.

Murray, CLJ, 1999. Health Inequalities and Social Group Differences: What Should We Measure? *Bulletin of the World Health Organisation*, 77: 537-543.

Shkolnikov Vladmir M., Tapani VA Begun and Evgueni M. Andreev, 2001. Measuring Inter-group Inequalities in Length of life. *GENUS*, LVII (n. 3-4): 33-62.

SRS, 2005. *SRS Based Abridged Life Tables, 1998-2002*. Analytical Studies Report No.1. Office of the Registrar General, India.

Wagstaff, A, 1991. On the Measurement of Inequalities in Health. *Social Science and Medicine*, 33: 545-557.

WHO, 1980. The Inequality of Death: Assessing Socio-economic Influences on Mortality. *WHO Chronicle*, 34: 9-15.

10

Gender Critique of Political Change in India

Miss. Sailaja Nandini

Introduction

Despite their vast numerical strength, women occupy a marginalized position in society because of several socio-economic constraints. This has inhibited effective participation of women in political processes and the institutional structure of democracy. According to the Document on Women's development (1985) women's role in political structure had virtually remained unchanged; despite the rapid growth of informal political activity by them. Broad based political participation of women has been severely limited due to the nexus of traditional factors such as considerations of caste, religion, and feudal and family status. As a result, women were left on the periphery of political process.

Historical Background

Taking the historical development of women's participation into account, in the first three decades of the history of the Indian National Congress women's participation in its activities was peripheral and negligible. Perhaps in those early days women went to those congress sessions more as the helpmates of their husbands, to see after their food and creature comforts. The agitation against the partition of Bengal and the Swadeshi movement attracted the attention of women in many parts of India. As a result, women started participating in the national movement actively. They took part in the Satyagraha campaigns, took out processions defying the ban on public meetings and processions and organized demonstrations in front of liquor shops and shops selling imported cloth.

Women Since Independence

With the dawn of Independence and the adoption of the Republican Constitution, several legal measures were taken by Parliament and the Government of India, which improved the status of women in India. The constitution guaranteed all women equal rights of participation in the political process of the country along with equal opportunity and rights in education and employment. The right to vote was guaranteed to men and women above 21 years age. The principle for equal pay for equal work was also recognized by the constitution which does not prevent the government from passing separate Acts or making special provisions in the acts for protecting the interests of women and children. The constitution also provides for right to an adequate means of livelihood for men and women equally (Article 39). During 1954-56, special marriage act, Hindu minority and guardianship act and Hindu adoption and maintenance act, Hindu Succession Act were passed which legalized intercaste marriage, ensured monogamy and conceded the right to divorce and equal share of property to Hindu women making women self confident to stand equal to the men in the family and society.

Women in Indian Politics

India was one of the first countries in the world to have a woman prime minister Indira Gandhi. She was a fiery politician, very dynamic, absolutely courageous, could hold her own in any platform in the world, and led India to a victory against Pakistan in the Indo-Pak war over East Pakistan, which resulted in Bangladesh becoming another country. But she had her own negatives—the worst of which was imposing a national state of Emergency in India in 1975 after she had lost a court case over irregularities during her election campaign in the previous elections to the Indian Parliament. The 19 months which followed, during which the press was gagged and censorship imposed, all her political opponents were put in prison, all civil rights suspended, and the like, are considered one of the darkest periods in the history of independent India.

Apart from her, several women politicians have made it to the position of chief ministers in a few states of India. Women parliamentarians are becoming more common, as also women legislators in our state Assemblies. But the fact remains that most of these women, including Indira Gandhi, reached the top positions because of their family history; they were, almost all, daughters, sisters, wives, mistresses, etc, of very prominent male politicians.

Women activists in India have been demanding a 33 per cent quota in Parliament for years, but this is not likely to materialise for a long, long time because the men are not ready to give up one third seats to women. However, at the grassroots level and at local body (Panchayat) levels, the 33 per cent reservation is already on and this is bringing around a silent revolution in many parts of rural India. Women chiefs of local bodies have made a big difference in constructing schools and health care facilities, improving infrastructure—roads, electricity, etc.—and are accessible to their rural constitutes to solve their problems. This holds out much promise for the future. But some women politicians at the top have become notorious in attracting bigger charges of corruption, arrogance and nepotism. And this is distressing.

Women and Electoral Politics in India

The concern for women's political equality in India first emerged as a political issue during the national movement in which women were active participants. As early as 1917, Indian women raised the issue of representation in politics, which at the time meant a demand for universal adult franchise. By 1929 women had the right to vote on the basis of wifehood, property and education. Under the Government of India Act 1935, all women over 21 could vote provided they fulfilled the conditions of property and education. Post-independence, newly adopted Constitution of India guaranteed equality and liberty, thereby enabling women to participate in electoral politics on an equal basis with men. They also continued to play a significant role in less conventional political activities such as environmental movement, anti-alcohol agitation, peace movement and even revolutionary activities.

Yet, politics proved to be a very inhospitable terrain for women and continues to be a male bastion into which the entry of women is severely restricted. Women's participation in politics remained in the shape on an inverted pyramid with larger percentages of women at the top and less at the bottom. The situation has been rectified somewhat by the 73rd and 74th Amendments but women continue to remain invisible and marginalized in decision-making bodies, leading to lack of a feminist perspective in political decision-making. Very few women have been successful in subverting the boundaries of gender and in operating in a very aggressive male-dominated sphere.

What does seem to be the case is that-barring striking exceptions where dynastic charisma is seen to matter more than anything else—most women politicians have found it difficult to rise within party hierarchies, and have managed to achieve clear leadership only when they have effectively broken out and set up parties on their own. Jayalalitha and Mamata Banerjee are clear examples of this, but there are other less well-known instances as well. Yet once these women become established as leaders, another peculiarly Indian characteristic seems to dominate—that is the unquestioning acceptance by the (largely male) party rank and file of the leader's decisions.

What all this suggests, therefore, is that the political empowerment of women not only still has a long way to go, but finally may not have all that much to do with the periodic carnivals of Indian electoral democracy. This is not to say that the electoral representation of women is unimportant, but rather that it needs to be both deeper and wider than its current manifestation in the form of the prominence of a few conspicuous women leaders.

Women's Political Participation and Leadership in the Governance of Municipal Institutions

India is a land of multiple cultures, religions, castes, languages and life styles. It has one of the largest histories of civilization. The constitution of India grants right to equality or equality before law; prohibition of discrimination on grounds of religion, race, caste, sex or place of birth and equality of opportunity in matters of public employment which includes equality to women also. During the last decade of twentieth century, India has witnessed significant changes in its constitutional and political setup. Before 73rd and 74th amendments in the constitution of India, the women and the weaker sections of the society were not getting adequate representation in the governance of local-government institutions in the country. The 73rd and 74th the Constitutional amendment have granted enhanced participation of young men and specially women in the urban and rural local-government institutions of the country. Now 33 per cent seats/constituencies in the local-government bodies are reserved for women and the young men and women of the above-referred weaker sections have also been given reservation in proportion to their population in the area.

From Annie Besant to Sonia Gandhi: The Changing Place of Women of Foreign Origin in Indian Politics

From Annie Besant to Sonia Gandhi, the Indian political history and politics has witnessed a significant change from its earlier position of a tolerant India of accepting and accommodating people of foreign origin particularly women in her political and social system to playing politics on the issue of birth origin. It also reflects alarmingly the growing feeling of intolerance and fundamentalism in secular Indian Politics. When Annie Besant, a noted British born activist of women rights came to India in November 1893 to do service to the Indian masses through Theosophical society nobody raised question towards her being foreigner. Even when she shifted from her spiritualism to direct politics of 'Home Rule' movement in 1916 against unjust British colonial rule nobody questioned her that, why a foreigner is involved in Indian political questions? Rather than criticism of being a foreigner, she was awarded for her genuine efforts for uniting moderate and militant factions of Indian National Congress Party as its party president in 1917.

But when Mrs. Sonia Gandhi, an Italy born wife of India's Late Prime Minister Mr. Rajiv Gandhi, took the Presidentship of Congress party in 1999, questions were raised from different quarters and also within the party of her being a foreigner involved in shaping country's destiny. Although being a "BAHU" (daughter in law) of Indira family, despite being a foreigner, she still enjoys mass popularity in rural and urban India. It also raises the gender and social question of respect of marriage institution and its influence in Indian politics to examine. So it becomes important to evaluate and analyze, why this change of perception took place on this issue among some political parties particularly present ruling Bhartiya Janta Party (BJP) and Nationalist Congress Party (NCP) on this issue of "origin". And to examine its relevance in the growth of 'fundamentalism' and 'intolerance' in contemporary Indian politics in the background of India's political history.

Women's Movement in India

The women's movement in India today is a rich and vibrant movement, which has spread to various parts of the country. It is often said that there is no one single cohesive movement in the country, but a number of fragmented campaigns. Activists see this as one of the strengths of the movement, which takes different forms in different parts. While the movement may be scattered all over India, they feel it is nonetheless a strong and plural force.

It is important to recognize that for a country of India's magnitude, change in male-female relations and the kinds of issues the women's movement is focusing on will not come easy. For every step the movement takes forward, there will be a possible backlash, a possible regression. And it is this that makes for the contradictions, this that makes it possible for there to be women who can aspire to, and attain the highest political office in the country, and for women to continue to have to confront patriarchy within the home, in the workplace, throughout their lives. As activists never tired of repeating: out of the deepest repression is born the greatest resistance.

Fifty years ago when India became independent, it was widely acknowledged that men had fought the battle for freedom as much by women as. One of the methods M K Gandhi chose to undermine the authority of the British was for Indians to defy the law, which made it illegal for them to make salt. At the time, salt making was a monopoly and earned considerable revenues for the British. Gandhi began his campaign by going on a march–the salt march – through many villages, leading finally to the sea, where he and others broke the law by making salt. Gandhi in his chosen number of marchers had included no woman. But nationalist women protested, and they forced him to allow them to participate.

The first to join was Sarojini Naidu, who went on to become the first woman President of the Indian National Congress in 1925. Her presence was a signal for hundreds of other women to join, and eventually the salt protest was made successful by the many women who not only made salt, but also sat openly in marketplaces selling, and indeed, buying it. Sarojini Naidu's spirit lives on in thousands of Indian women today. Some years ago, Rojamma, a poor woman from the southern state of Andhra Pradesh, attended a literacy class. Here, she read a story which described a life very like her own. It talked about a poor woman, struggling to make ends meet, whom did her husband regularly beat. Whatever he earned, he spent on liquor, and then, drunk and violent, he attacked her because she had no food to give him. Unable to stand the continuing violence, the woman went from house to house, to find every other woman who had the same story to tell. They got together, and decided they would pitch their attack where it hurt most: they would picket liquor shops and stop liquor being sold. Their husbands then would have no liquor to drink, and the money they earned would be saved. Inspired by the story, Rojamma collected her friends together, and they began to picket liquor shops. The campaign spread like wildfire. In village after village, women got together, they talked, they went on strike, they beat up liquor shop owners, and they refused to allow their husbands to squander money on liquor. And, they succeeded. The government banned the sale of liquor in Andhra Pradesh, reluctantly, for liquor brings in huge amounts of money. As a result, savings went up, violence levels dropped, and the lives of poor women began to improve.

The hundreds of thousands of Rojammas and Sarojini Naidus who are to be found all over India form part of one of the most dynamic and vibrant of political movements in India today, the women's movement. The trajectory of this movement is usually traced from the social reform movements of the 19th century when men took up, campaigns for the betterment of the conditions of women's lives initially. By the end of the century women had begun to organise themselves and gradually they took up a number of causes such as education, the conditions of women's work and so on. It was in the early part of the 20th century that women's organisations were set up, and many of the women who were active in these later became involved in the freedom movement.

Independence brought many promises and dreams for women in India–the dream of an egalitarian, just, democratic society in which both men and women would have a voice. The reality, when it began to sink in was, however, somewhat different. For all that had happened was that, despite some improvements in the status of women, patriarchy had simply taken on new and different forms. By the 1960s it was clear that many of the promises of Independence were still unfulfilled. It was thus that the 1960s and 1970s saw a spate of movements in which women took part: campaigns against rising prices, movements for land rights, peasant movements. Women from different parts of the country came together to form groups both inside and outside political parties. Everywhere, in the different movements that were sweeping the country, women participated in large numbers. Everywhere, their participation resulted in transforming the movements from within.

Worried at this increase in political activity, Indira Gandhi's government declared a State of Emergency in 1975, putting a stop to all democratic political activity. Activists, both young and old, women and men, were forced to go underground or to stop all political work. It was only when the Emergency was lifted, some 18 months later, over ground political activity resumed. It was around this time that many of the contemporary women's groups began to get formed, with their members often being women with a history of involvement in other political movements.

One of the first issues to receive countrywide attention from women's groups was violence against women, specifically in the form of rape, and what came to be known in India as 'dowry deaths'–the killing of young married women for the 'dowry' or money/goods they brought with them at marriage. This was also the beginning of a process of learning for women: most protests were directed at the State. Because women were able to mobilize support, the State responded, seemingly positively, by changing the law on rape and dowry, making both more stringent. This seemed, at the time, like a great victory. It was only later that the knowledge began to sink in that mere changes in the law meant little, unless there was a will and a machinery to implement these. And that the root of the problem of discrimination against women lay not only in the law, or with the State, but was much more widespread.

In the early campaigns, groups learnt from day to day that targeting the State was not enough and that victims also needed support. So a further level of work was needed: awareness raising or conscientisation so that violence against women could be prevented, rather than only dealt with after it had happened. Legal aid and counseling centers were set up, and attempts were made to establish women's shelters. It was only when groups began to feel sucked into the overwhelming volume of the day-to-day work of such centers that they began to feel that it was not enough to do what they now saw as 'reformist' and 'non-campaign' work. Knowledge was recognized as an important need. India is such a vast country; what did activists in Karnataka, a state in southern India, know of what was going on in Garhwal in north India? And yet, everywhere you looked, there was women's activity, activity that could not necessarily be defined as 'feminist', but that was, nonetheless, geared towards improving the conditions of women's lives.

In recent years, the euphoria of the 1970s and early 1980s, symbolized by street-level protests, campaigns in which groups mobilized at a national level, the sense of a commonality of experience cutting across class, caste, region and religion – all this seems to have gone, replaced by a more considered and complex response to issues. In many parts of India, women are no longer to be seen out on the streets protesting about this or that form of injustice. This apparent lack of a visible movement has led to the accusation that the women's movement is dead or dying.

Perhaps the most significant development for women in the last few decades has been the introduction of 33 per cent reservation for women in local, village-level elections. In the early days, when this move was introduced, there was considerable skepticism. How will women cope? Are they equipped to be leaders? Will this mean any real change, or will it merely mean that the men will take a backseat and use the women as a front to implement what they want? While all these problems still remain, in a greater or lesser degree, what is also true is that more and more women have shown that once they have power, they are able to use it, to the benefit of society in general and women in particular.

Women in Local Governance

With the PRIs getting constitutional status by way of the 73rd Amendment Act, it is hoped that the women will have a greater participation in the PRIs.The 73rd Amendment Act provides for reservation of one third of seats for women and these seats may be allotted by rotation to different constituencies in Panchayats.The participation of women in the PRIs is considered essential not only for political participation in the democratic process but also for realizing the developmental goals for women. Though women are coming, it has been observed that women find it easier to enter into politics at local level than at the national level, the reason being that the entry eligibilities are liberal. Although women are elected in large numbers, their participation has been full of challenges and obstructions. Women's interests in politics do not matter and their husbands, fathers and brothers to carry out mandates use them as puppets.

The Indian government has also pursued political empowerment of women and now in Rajasthan for example there are 3000 women village heads or Sarpanches and 33000 elected women representatives on Panchayat. At the central level the 73rd Constitutional Amendment, 1992 includes a provision for statutory minimum reservation of 33 per cent seats for women in Panchayati Raj. India has had women Prime Ministers and Governors and now Indian women are entering jobs, which were previously only reserved for men. Women in the fields of medicine and computer technologies have made phenomenal progress. Despite such progress crimes against women has been increasing over the years.

Women's Reservation Bill—A Critique

The real test of democracy is the creation of equality of opportunity for the hitherto deprived sections of society. It requires both a favourable social atmosphere and an individual attitude. Individual attitude and social atmosphere is a sort of reversible equation: one influences the other, in both directions. In practical terms it means that efforts have to be made at various levels of society simultaneously. Every attempt, in every direction, is bound to affect adversely some vested interests. So, one has to be prepared for a long drawn out struggle on all the fronts. Democracy in kitchen and bedroom goes hand in hand with democracy in Parliament and Panchayat. It has to become a way of life; it has to be adopted in literary vocabulary and in political discourse alike.

In the context of the present discussion it amounts to shedding of all mental reservations against reservation of seats for women in the Parliament and in Assemblies. The idea of making a legal provision for reserving seats for women in the Parliament and State Assemblies came into being during Rajeev Gandhi's tenure as the Prime Minister of India when the Panchayati Raj Act, 1992 (73rd and 74th Constitutional Amendment) came into effect granting not less then 33 per cent reservation to women in the Panchayati Raj Institutions or local bodies. Prime Minister H.D. Deve Gowda made the actual promise for reservation of seats for women in Parliament and State Assemblies in 1996. I.K. Gujral proposed the present form and shape of the Bill during his term as the Prime Minister of India.

The Bill in its Current form envisages reserving 181 seats in the Parliament for women. In practical terms its efforts would be that 181 male members of Parliament would not be able to contest elections if the Bill is passed. Also, there is to be a rotation of seats, i.e., a male member of Parliament can not represent the same constituency for more then two consecutive terms. Here lies the rub.

These two very provisions are seemingly the cause of the consensus arrived at by various political parties to dump the Bill. 181 seats in Parliament is too great a number to be sacrificed for the mere ideal of women's empowerment or adequate political representation, the very idea makes the male politicians panicky. The clause of 'rotation of seats' is seen by the opponents of Bill to 'strike at the very heart of democracy and democratic values' as, according to their logic, the representative will not get a chance to nurture his constituency nor the electorate will get a chance to reward or punish their representative, as a corollary to it hardly any ties would be established between the two.

This argument may hold water when it is discussed in classroom sessions but it cannot be taken as the sole basis to discard Women's Reservation Bill altogether. Securing 33 per cent reservation for women in opening the doors of opportunity for political empowerment to almost 50 per cent of our population. It will not only serve the cause of democracy as the Panchayati Raj Institutions are doing at the grassroots level but will also go a long way in ensuring political equality through active participation of women from both urban and rural areas. Also, if social equality through political empowerment is to be achieved, the bill should include clauses, which guarantee quota within quota to women belonging to scheduled tribes, scheduled castes, other backward castes, and minority communities so that a level playing field is provided for them as well.

It is also argued that the Bill in its present form would end up ensuring seats in Parliament for the female relatives of those who are already in power. To counter this situation, provisions can be added in the Bill, which provides for no reservation to women who have close relatives in active politics (an acceptable definition of 'close relatives' can easily be arrived at.) These women can contest from general seats. There had been suggestions in the past in the form of alternatives to the Bill. One is to amend the Representation of People's Act 1951, to compel political parties to nominate women for one-third of their seats or lose recognition. This, according to Rajindar Sachar, former Chief Justice of Delhi, is flawed, as it would violate the Constitution of India, which guarantees its citizens the right to form association under Article 19(1)(c) as a fundamental right. Another alternative is to increase the number of seats in the Lok Sabha, which is currently based on the figures of the census of India, 1971, when the population of India was 54 crore. The number of seats were limited to 530 till further amendments. Now the Delimitation Commission has been asked to take the 2001 census as the basis for delimiting constituencies. According to 2001 census, the population of India has risen to 102 crore; therefore the number of seats are bound to increase before the next general elections. This should be reason enough to pave the way for the safe passage of the Women's Reservation Bill.

Moreover, when the so called backward and fundamentalist society like Pakistan can grant 33% reservation to women in its Senate then why should India, the largest democracy in the world, lag behind?

Conclusion

Despite a long standing and vigorous women's movement with many achievements, patriarchy remains deeply entrenched in India influencing political and social institutions and determining opportunities available to men and women. There is an urgent need to redefine feminist political agency—to allow for the possibility of secular political collectives to which women can belong not by ascription, but by voluntary participation. In other words, to rebuild the fragmented constituency of women. But the solution may not lie in a revitalized national women's movement, particularly in the current context of economic fragility and political instability. There is already the contour of a women's movement, actual or potential, in the impressive networking capacities of autonomous groups and the mobilizing potential of left-led women's groups. We need to recognize the importance of women's associations at the local and regional levels without retreating into the irreducibly local. The developing and strengthening of local institutions is also necessary, and seems to dominate women's movements in the late 1990s. The question is how to mobilize these institutions for a transformative feminist politics—that is, to ensure that these localized struggles face and accommodate the challenges posed by community and caste politics without allowing them to displace gender concerns.

REFERENCES

Dasgupta, K(1998). *"Reservation for Women's Representation"*, INSPARC, Kalyani.

Madhu Kishwar(2000) *Equality of Opportunities vs. Equality of Results, Improving Women's Reservation Bill*, Economic and Political Weekly, Vol. XXXV, No. 47, November 18, 2000, pp. 4151-4156.

Meena Dhanda (2000) *Representation for Women, Should Feminists Support Quotas*, Economic and Political Weekly, Vol. XXV, No. 33, August 12, pp. 2969-2976.

Nivedita Menon (1999) *Gender and Politics in India* Edited by Nivedita Menon, Delhi, Oxford University Press.

Omvedt, Gail. (1990) *Violence Against Women: New Movements and New Theories in India*. New Delhi: Kali for Women.

Rajan (1999) *Signposts: Gender Issues in Post-independence India*. Edited by Rajeswari Sunder Rajan. New Delhi: Kali for Women.

Ray, Raka. (1999)*Fields of Protest: Women's Movements in India*. Minneapolis, Minn.; University of Minnesota Press.

Sangari and Shimla (1999) *From Myths to Markets: Essays on Gender*. Edited by Kumkum Sangari, Uma Chakravarti. Shimla: Indian Institute of Advanced Study; New Delhi: Manohar Publishers and Distributors.

The History of Doing, An Illustrated Account of Movements for Women's Rights and Feminism in India, 1800-1990, Radha Kumar, pp. 197.

Varma, Sudhir(1997) *Women's Struggle for Political Space: From Enfranchisement to Participation*. Jaipur: Rawat Publications, 1997.

"World Conference on Women—An Indian Perspective", (1995) Mainstream, September.

11

Role of Women on the Political Life of Orissa

Mr. Santosh Kumar Sahu

From 'no voting rights' to 'participation in governance and polity' has been a long and arduous journey for women across the globe. The process began in the early part of 20th century and over a period of nearly hundred years the participation of women in Parliament and Legislatures of the states has reached eleven per cent of the total seats. Political participation is the hallmark of a democratic set up. Nature, success and effectiveness of a democracy largely depend upon the extent to which equal, effective and actual participation is provided by the system to all its citizens. A democracy will fail in its objective if a vast number of women lack equal opportunity to participate in the governmental decision making process. As women comprise nearly half of the population, this segment of society cannot be ignored. Women are also equal partners in nation building and political development.

Participation provides the individual with an opportunity to express views on important issues and influence Governmental polices in desired direction.

In State Government, the Legislative Assembly is supreme in its scope of power and authority. Since it is composed of elected representatives of the people, it can be called as the mirror of the state. The will and wishes of the state are therefore reflected through its elected representatives inside the legislature. We see some women leaders coming to power by their own effort. The women elite have played significant role in converting the inputs into political outputs at the society. These women legislators try to bridge the gap between the people and Government. As John Wahlke and Heinz Eulau, observe "of all political institutions, none is more vital to the process of linking governors and the governed in relationship of authority, responsibility and legitimacy" (Mishra, 1991). They have always tried to develop the infrastructure of their constituency and state whenever they come to power polity as MLA and MP.

Participation of women in polity from Panchayati Raj to Assembly and Parliament, operating in bureaucratic structure is an important feature of women awakening in the post-independence period in Orissa (Samal, 1996). Participation of women in the Governance can be traced through hierarchy of political involvement. Such system consists of a series of upward moving levels that are:

(i) Voting

(ii) Participation in grass roots-democracy.

(iii) Women contestants in Assembly and Parliamentary Election

(iv) Ministerial Responsibility

Women representation has been noticed in the composition of all ministries, since Orissa achieved statehood on 1st April 1936. Women in Orissa have their first socialization in the political sphere during the struggle for independence. After Orissa became an independent state within British India, women availed greater opportunities to particpate in reconstructed society and legislative polity (Tripathy, 1991). There has been a steady increase in political participation of women after India became Independent in 1947, as women attained bonafide citizenship of a new nation. The first general election was held in 1952 on the basis of adult franchise and since then women have continued to participate in state Legislative and Parliamentary election (GOO, 1953).

Election opens up channels between polity and society, between individuals and the Government. It is a major agency of political individual and the Government, also major agency of political socialization and participation. Therefore, the role of women in this field can be measured by their voting behaviour, participation in grassroots democracy, contesting elections and occupying seats in legislative assembly and in parliament, then the performance of responsibility.

Voting is the basic activity by which the citizens get assimilated in the political process. It has a tremendous impact for equalizing and mobilizing women. Political behaviour of persons are fully exposed during the time of election when men and women as voters exhibit various trends of thought and mind. Political Scientists, therefore make a study of behaviour of voters after every election, which becomes very useful for drawing conclusions on the functioning of various institutions (OPSJ, 1991). As electors women's role is substantive and compares favourably with the male voting turn-out. The voting participation of the women was small in number. In 1952 and in 1980 election the proportion of women voters was only between ten and eleven percentage points behind that of men voters. However, nearly 55 per cent women voted in 1985 elections as against 47 per cent in earlier elections. In 1989-90 Assembly elections 55 per cent to 60 per cent women voters exercised their voting right. It was seen that the percentage of poll for women in state crossed the national average. In 21st century with comparison to earlier periods, women voters seem to be politically very much conscious. Further, the study of past certainly gives much political weightage to the women voters of Orissa for their rich experience of various political movements and personalities. In this context the fact that enlightened women voters of Orissa have keen interest in local, regional, national and international polity is certainly a challenging phenomenon.

Women participation in grass-roots democracy assumes a vital significance in the context of women's viability in the local decision making process, for the majority of women in rural areas belong to the lower strata of the society. This participation and leadership in rural political institutions can be of immense help not merely towards their empowerment and development but also it brings to bear a solid feminine perspective to the process of planning, policy formulation and execution of rural development programmes. The report of the Committee on the Status of Women published in 1974 pleaded for provision of special opportunity for women, representing in local Government. In 1978 the Committee on Panchayati Raj Institutions recommended the Reservation of seats in Panchayats.

The 73rd and 74th Amendments of 1993, to the Constitution of India have provided for reservation of seats in the local bodies of Panchayats and Municipalities for women, laying a strong foundation for their participation in decision making at the local levels (GOI, 1993). These amendments provide reservation for women, one-third of seats in Grama Panchayat, Panchayat Samiti and Zilla Parisad. The role of women at all the three tires is significant. The amendment has empowered SC and ST women also in PRIs. Before this amendment also women were participating in Panchayati Raj Institution but the number increased afterwards. Any categories whether General or ST or SC should come forward to take the responsibility of local area.

Throughout the history of human civilization, right from ancient to modern times, women have always been an oppressed lot. They have faced social barriers and prejudices in all spheres of activities. Their contribution to building up a strong society by providing love, care, sacrifice and instilling good human principles in their children has always been considered to be of minimal value. That is the reason why, although civilization has progressed to a great extent, the majority of women all over the country are still relegated to the background. Women of Orissa too have the same situation as their sisters all over the country. These large majorities of women are still suffering silently within the four walls of their houses. However, it is a matter of pride, that in spite of stiff opposition and oppression from society, a good number of women are succeeding and emerging from below and ensuring their place in the history of Orissa.

At the outset it is to be noted that constitutionally and legally there is no bar on women's participation in polity. But a very few women are actively involved in political sphere. An enormous disparity exists between women's formal political equality and their meaningful exercise of political power. They exercise their franchise in large numbers, but when it comes to enjoying power position or occupying prestigious political offices, they lag behind. The strength of their voting number is nowhere reflected in their direct role in the Government. This situation is not peculiar to Orissa only.

However political roles are shaped by the social millieu in which people live. In ancient Orissa women enjoyed a position of equality in the family and society; hence, they also participated in political activities along with men. We find some evidence of women's active participation in public affairs during the Mauryan and Bhauma period too, who had at the hour of need, led the armies, directed the government and ruled the kingdoms and enjoyed considerable freedom and privileges in all spheres of family and public life. In the royal courts female were engaged not only as chauri bearers, parasol bearers but also appointed as doorkeepers, guards and spies. Shapping the history of ancient Orissa so many women played significant role as we have seen. Equal importance was given to women in Kharavela's time. The chief queen of Kharavela at times used to help in the administration and as such she commanded great respect and influence in the realm. From the Manchapuri cave inscriptions we know that women in this period were highly accomplished, not only with dancing and playing of various instruments but also well acquainted with nursing and military exercise, such as driving elephants and hunting. Princess of Kalinga, helped their fathers in keeping good terms with other kings.

In course of time there was a gradual decline in their position due to certain internal changes in the society. Institution of marriage, lack of education and some Brahmanical notions added to their ignorance and made them dependent upon men folk, in all walks of life including political affairs. However, the Muslim occupation of Orissa during the 16th century affected the life of women. We do not find accounts of any great Rajput, Muslim and Maratha women in the history of medieval Orissa. They were deprived of their power and position. Child marriages, prohibition of female education and practice of purdah confined them to the four walls of home and check was imposed upon their public life.

However, social reformation and women's education in the late 19th and early 20th century gave rise to a new consciousness among women. A number of women from all sections of society moved by patriotism crossed the social barriers and joined the national stream in Orissa. Women were not so much politically active in Orissa prior to the year 1921.

It was with the advent of Gandhi in the political scenario of India, there was a remarkable change marked among women. Gandhis's first visit to Orissa in March 1921 marked the beginning of political awakening among women of Orissa. Though the number of women who joined the non-cooperation movement was small, yet a beginning was made. Subsequently women in large numbers joined the various programmes of the National Congress and very effectively played their role, till the attainment of independence. In the political movements of the period, they participated as volunteers, active workers, organizers as well as leaders. They organized meetings, delivered speeches, led procession, violated government law, defied government orders, composed and sang patriotic songs, unfurled and hoisted Congress flag, faced police lathi charge and atrocities and courted arrest. They actively participated in the Non-Corporation, Civil Disobedience and Quit India Movement and achieved independence.

With the advent of Independence, women were guaranteed political equality with men. In consonance with the right of universal adult franchise and all other political rights conferred on them by the constitution, women's participation in political activities increased. With the process of modernization and spread of education their mental horizon broadened, a change took place in their social position and they came out of the four walls of homes. Consequently, they started taking active part in polity and public life.

However, the overall situation of Orissa in regard to women's participation in polity has not been very satisfactory. Barring a few exceptions, women have remained outside the domain of power and polity. In this state, although the number of women voters has been nearly equal to the number of male voters, sometimes even more, yet their representation in the state legislature, parliament and decision making bodies has been small in number. They are at a great disadvantage in power politics. A very small percentage of women are put up by different political parties to occupy important positions in the party hierarchy or as candidates in the election. Actual power is male monopoly with just an electoral plank. No party met the target of 30 per cent reservation for women in accordance with the national perspective plan. However, the actualization of 30 per cent reservation at the grassroots level has opened up new horizons for women's participation in the state polity. Further, there has been a rapid rise in unconventional political activities by women including protest movements. Though with course of time a change has taken place in society and women's status, their participation in political affairs still remains marginal. They are nowhere near men in political matters.

Women have not entered the political life by choice due to reservation policy, the politically influential persons or local leaders persuaded the female members of their families to contest for the posts. Hence most of the elected women in active polity come from politically active families and belong to well-to-do families. Moreover, women families having better economic status participate more in polity than those having low economic status. Only few women, who are elected in grassroots level are economically backward but have won the support of their local people.

Membership of women in different political parties is confined mainly to the lower level that is village and block level, very few women make it to the higher levels, state and national. Very few women contest for the state Legislative Assembly and a insignificant number of them contest for the parliament. The influential leaders openly and latently dominate the polity and they do not allow educated and confident women get elected in this sphere, who could challenge their power. They support such women representative (generally members of their family or caste) who would let them do everything on their behalf. Other members do not oppose either because of fear or because they themselves pursue their selfish ends. However, influence of male family member is also of great significance even in the case of educated women as wives, widows, sisters, daughters and daughter-in-laws of political leaders. In short, most of the women representatives are like puppets in the hands of these influential persons. The males who are behind these women representatives pull the strings. They also accompany these women in the meetings to direct and control them. Generally they themselves participate in discussions and decision-making process. Mostly these women representatives sit in the meetings as passive participants. The number of women leadership positions at the local, village, district and national level is still not commensurate with their numbers in society.

In 21st century a change has taken place in women's perception of their role. They feel that their role in society is not confined to only bearing and rearing children or to family responsibilities. They do not accept polity as the exclusive sphere of men. Some of them contest elections of their own choice and will. At the same time, they do not seem to be satisfied with the present representation of women in a Panchayats, the state legislative assembly and the Parliament and want more a representation of women in various decision making bodies. The vital question of under representation or invisibility of women in decision-making process, their deprivation leading to unequal distribution of resources, neglect of their interests, needs proper debate and discussion to ensure national progress and development.

In fact there is no denying empowerment of women in all spheres, especially the political sphere is crucial for their advancement and the foundation of a gender-equal society is central to the achievement of the goals of equality development and peace.

Orissa women have come a long way in the paths of development since the days of independence. But their real participation in numerical terms is not enough to make their presence to such dampening state of affairs.

Women are reluctant to join polity because of lack of political knowledge lack of political interest and political education, unfavourable family and societal attitudes and also because of their pre-occupation with domestic responsibilities. Men like to retain their monopoly of power and do not want to encourage women to join polity. Socio-economic marginalization has driven them to the backwater polities. Women's lack of access to financial resources is a strong deterrent to their effective participation. Women rarely have the energy or the inclination to get out of their homes to realize their political ambitions. Political parties direct their efforts to mobilizing women mainly around elections and during election time overlooking the advantages in systematic mobilization of women voters around key issues. Lastly illiteracy inequality of class, status and power, cultural restraints, socialization by a patriarchal society have put women as an insignificant minority in the political domain. Though state has made efforts to improve opportunities for women's participation in political life the economic social and cultural barriers continue to limit their participation.

There is no denying the fact that if women are given right type of orientation and opportunities then they can play a meaningful role in the society. But because of low representation most of the crucial issues related to women's development are ignored unattended and uncared for.

Realizing that it is necessary to elevate the status of women the state and political parties in general should cooperate to achieve the success. The parameters which can help women to join in the decision making process of state are:

- To boost their self-confidence and assertiveness, so that they can take steps to becomes self reliant. They have to build a position of self-image.
- Education of women occupies top priority amongst various measures taken to improve the status of women in the state. So efforts are to be made to improve the female enrolment in educational institutions.
- Family members should also co-operate with women by sharing their family responsibilities and encouraging them to take part in political activities.

- Women have to be economically independent. They must take advantage of all the economic opportunities; then only they can be free to choose their career in polity.
- Women in themselves should develop a favourable attitude towards polity. They should spare time and come out of their homes to take part in political activities.
- The state government must provide certain protective and safety measures to women.
- Above all, there is a need to replace the traditional value system, which is based upon inequality of sexes and in which women play a subordinate role. Therefore, in the process of bringing about social change, equal participation must be ensured by the state in decision-making at planning and policy level to reduce inequalities and unjustifiable disparities.

The participation of women in Governance and Polity can be seen in Parliament, in the State Legislative Assemblies and in other policy and decision making bodies. Due to higher education, self-awareness, taking employment in various field and national consciousness, they are now claiming equal status with men both inside and outside the family. They are not demanding superior status but equal status which is a true policy in a democratic setup. They are passing through a transitional stage, neither wholly traditional nor fully modern. Cultural heritage and traditional values cannot be wholly vanished from the attitudes and behaviour of women. On the whole the result will indicate a positive change. The political empowerment of women is critical not only for their development and for changing their status, but also to their emergence as equal partners in the progress of state and nation, Political participation and empowerment is not an end in itself, but a means to realize their creative energy for their economic independence and socio-political identity as equals.

Let us hope that just like women of past, today's women also rise up to do their best for society, state and nation, let the whole world know the dexterity of women of Orissa.

REFERENCES

S.K.Mishra, "The Orissa Legislative Assembly as the Guardian of the Weaker Sections of the Society", OPSJ, Vol. IX, 1990-91, p. 61.

J.K.Samal, "A Study of the Social Development of the Women of Orissa in the Post Independent Period, 1947-91", The Journal of Orissa History, Vol. XV, 1996, p. 95.

S. Tripathy, "Participation of Women in the Electoral Process of Orissa", OPSJ, Vol. IX, 1990-91, p. 45.

Report in 1st General Election in India, 1951-52, Vol. I, Govt. of Orissa, 1953, p. 77.

OPSJ, Vol. 1, 1990-91, p. 71.

J.K.Samal, op.cit, p. 96.

National Policy for the Empowerment of Women 2001, Department of Women and Child Development, Ministry of HRD, Govt. of India, p. 4.

Zenab Banu, "Tribal Women Empowerment and Gender Issues", Grassroot Panchayatraj: The Tribal Women, 2001, p. 105.

Orissa Census Report, 1971-2001.

India Census Report, 1961-2001.

Newspaper Collections.

12

Some Instances of Gender Disparity in Orissa

Mr. Akshaya Kumar Panigrahi

Introduction

The initial years of planning process in India were based on the trickle down theory on the belief that fruits of development and growth will spread out equally among all regions and all sections of society. But this did not happen as a large segment of our population especially those in weaker sections and women have been left outside the growth process. Even among women, those in rural areas are more disadvantaged as compared to those in urban areas.

Traditionally women bear primary responsibility for the well being of their families. Yet they are discriminated systematically and denied access to resources such as education, health services, employment, freedom etc. (Soundari and Sudhir, 2003).

According to Amartya Sen, gender inequality is not one homogenous phenomenon, but a collection of disparate and interlinked problems and the different kinds of gender inequality are displayed in mortality, natality, basic facility, special opportunity, ownership and household.

Female education is not only an end itself but also serves as a catalyst for overall performance in other segment too (Soundari and Sudhir, 2003). Once a woman is educated then her children definitely get proper education. There should be policy for greater opportunities for education of women, which will create multiple linkages. Opportunities have to be enhanced for women so that they can generate income and be self-supportive. Women's empowerment must be accompanied by women's health. The patriarchal values need to change and public awareness that a nation cannot be healthy unless its women are healthy. Gender equality can bring efficiency in all economic sectors. But it's a long way to go. The journey has just begun. This requires concerted efforts at all levels to make this endeavour a success.

For various historic and socio-cultural reasons, women are a vulnerable section of our society and several macro indicators related to education, health, employment, economic participation etc. point towards an adverse status of women vis-à-vis; men, women also comprise a sizeable segment of the poverty struck population. Development and empowerment of women has been priority in successive plans and several expenditure programmes are directed to this objective, yet, women face gender specific barriers to access of public services and expenditure (Soundari and Sudhir, 2003; Kateja, 2001).

Understanding Gender

Gender refers to the different roles that men and women play in society and the relative power they wield. Gender roles vary from one country to another, but almost everywhere, women face disadvantages relative to men in social, economic and political spheres of life. Where men are viewed as the principal decision makers, women often had a subordinate position in negotiation about limiting family size, contraceptive use, managing family resources, protecting family health or seeking jobs. Gender differences affect women's health and well being throughout the life cycle. Following are some of the instances of gender disparity in the society:

- Those parents who prefer sons may put girls at risk of sex selective abortion or infanticide before or at birth;
- Where food is scarce, girls often eat last and usually less than boys;
- Girls may be less likely than boys to receive health care when they are ill;

- Adolescent girls may be pressurized into having sex at an early age, within an arranged marriage;
- Married women may be pressurized by husbands or families to have more children than they prefer and women may be unable to seek or use contraception;
- Married and unmarried women may be unable to deny sexual advances or persuade partners to use a condom, thereby exposing themselves to the risk of STIs;
- In all societies, women are more likely than men to experience domestic violence.

Cultural and social factors largely explain women's lack of power in protecting themselves on their daughters from these health threats. These factors include women's limited exposure to information and new ideas, ignorance of good health practices, limited physical mobility and lack of control over money and resources.

Gender disadvantages intertwine with poverty. Poverty is strongly linked to poor health, and women represent a disproportionate share of the poor. Women in the poorest households have the higher fertility, poorest nutrition and most limited access to skilled pregnancy and delivery care, which contribute to higher maternal and infant death rates. Women's low socio-economic status also makes them more vulnerable to physical and sexual abuse. Unequal power in sexual relationship exposes women to coerced sex, unwanted pregnancies and sexually transmitted infections. Impoverishment can also lead some women into commercial sex. Thus, women's access to and control over resources can give women greater control of their sexuality, which is fundamental to controlling their health.

Human development necessarily encompasses the issue of gender inequality. There are gender gaps in connection with rights, access, and control of resources in economic opportunities as well as for power and political representation. In no country do men and women have equal social, economic, and legal rights. Women still possess less of a range of productive resources, including land, education and financial resources (World Bank, 2000). Gender inequalities lead to higher levels of malnutrition, poverty, illness and other deprivations, with an adverse impact on the quality of life, productivity of farms and enterprises and governance.

Factors that lead to gender disparities in many countries include social norms, rights and laws, household decisions and the resulting resource allocations and economic policies that affect the level of household income and its distribution among household members (World Bank, 2000).

The 1994 International Conference on Population and Development (ICPD) articulated a bold new vision about the relationships between population, development and individual well being. At the ICPD, 179 countries adopted a 20 year forward looking Programme of Action (PoA), which built on the success of population, maternal health and family planning programmes of the previous decades while addressing, with a new perspective, the needs of the early years of the twenty first century.

The ICPD PoA defined a set of strategic objectives and spelled out corresponding actions to be taken by governments, the international community, non governmental organizations and the private sector to remove existing obstacles to gender equality and to improve the lives of girls and women. The Global Survey posed a number of questions on gender related issues in an attempt to track progress made by countries in implementing the PoA's gender component. The responses cover measures taken in five specific areas, namely: (a) protecting the rights of girls and women; (b) women's empowerment; (c) gender based violence (GBV); (d) gender based disparities in education and (e) men's support for women's rights and empowerment.

Similarly the Millennium Development Goals (MDGs) aims to promote gender equality and empower women, with fixing target to eliminate gender disparity in primary and secondary education preferably by 2005 and in all levels of education not later than 2015. Both the ICPD PoA and MDGs set targets for providing universal primary education, promoting gender equality and empowering women, reducing child mortality, improving maternal health and combating HIV/AIDS and other diseases. According the ICPD declaration it aims at 'Advancing gender equality and equity and the empowerment of women and the elimination of all kinds of violence against women, and ensuring women's ability to control their own fertility, are cornerstones of population and development related programmes. During the 1990s, global conferences sought to provide a vision of women's lives. The World Conference on Human Rights held in Vienna in 1993 asserted that women's rights are human rights. In Cairo, in 1994, the ICPD built on this assertion and placed women's rights, empowerment and health, including reproductive and sexual health, at the centre of population and sustainable development policies and programmes. At the Fourth World Conference on Women, held in Beijing in 1995, a consensus was reached that seeks to promote and protect the full enjoyment of all human rights and the fundamental freedoms of all women throughout their life cycle.

An important component of human condition is equity among races, genders, regions etc. Some sections of society are comparatively less benefited with the advancements recorded. One such segment is women. There is discrimination intended or unintended against women. They lag behind men in all fields including health, education, mobility, mortality and freedom etc. (Kateja, 2001). It is true in India, that the women lag behind the development process. Not only India, but also Orissa is not exceptional this scenario. Though the national indicators of women development is showing some standard number but in Orissa it is very much worst. So studying the gender disparity in Orissa especially on education and health is justified.

Demographic Profiles of Orissa

According to 2001 census there are 3,68,04,660 number of total population consisting 1,86,60,570 males and 1,81,44,090 females. Among them 85 per cent are from rural Orissa. The population is predominantly Hindu around 94.67 per cent. It has the third lowest population density (236 persons per sq. km. in 2001) among the major states of India, ahead of only Rajastan and Madhya Pradesh. However there is significant inter district variation within the state in this regard, with the district of Khurda having a population density of 666 persons per sq. km at one end and Kandhamal district with a population density of only 81 persons per sq. km at the other end. This has meant massive spatial concentration of the population. Coastal Orissa accounts for some 52 per cent of the population of the state with an area share of 25 per cent.

The rate of growth of population in Orissa during the decade 1991-2001 was 15.94 per cent as against 21.34 per cent for all India. This is the third lowest rate of growth of population among the major states of India with only Kerala (9.42 per cent) and Tamil Nadu (11.19 per cent) having lower rates. This has been the result of a peculiar demographic regime, relative low and steadily declining birth rate going hand in hand with relatively high and very slowly declining death rate, something that does not really fit into any of the three stages of the standard theory of demographic transition. The rate of urbanization in Orissa at 14.91 per cent is the lowest among the major states of India and is rising very slowly. But there is significant inter district variation in this respect, with the district Khurda in central coastal Orissa having an urbanization rate of 42.93 per cent at one end and Boudh in south central Orissa, having an urbanization rate of only 4.82 per cent at the other.

The life expectancy at birth in Orissa is around 60 years (61 years for males and 59 years for females). The sex ratio is favourable to women, which consist of 972 women per 1000 male population whereas at all India level it is only 933 women per 1000 male population. The literacy rate of Orissa is nearer to the national average. It constitutes the 63.61 per cent, whereas the all India literacy rate is 65 per cent. Similarly there is a lot of variation in demographic profiles of men and women in Orissa.

Gender and Education in Orissa

Education is both an indicator and an instrument of development, and its attainment is a major factor behind the accumulation of human capital. Education increases labour productivity in both urban and rural sectors and has long been identified as one of the most important determinants of economic growth. The gender disparity in Orissa's literacy rate in 2001 was 0.4901 as against 0.8192 in 1991, a decline of 0.3291 points. This disparity is found to be much less in the developed coastal districts such as Khurda, Cuttack and Jagatsinghpur, whereas it is quite high in the backward districts such as Kalahandi, Nuapada, and Nabarangpur, mostly in the KBK (Kalahandi, Bolangir and Koraput) region in Orissa. There is considerable variation in literacy rates across districts, especially amongst females. The coefficient of variation (CV) in literacy rates in 2001 among the districts is 23.95 per cent, with a value of 18.64 per cent for males and 32.51 per cent for females. But in 1991, the coefficient of variation in overall literacy rate was 30.88 per cent. Thus, though the gender and regional disparities in literacy rate have decreased between 2001 over 1991, the disparity is still high.

In order to reduce the gender disparity in schools, various strategies and activities have been adopted through the District Primary Education Project (DPEP) and the Sarva Sikshya Abhiyan (SSA). These include:

(i) Supply of free textbooks for girls;

(ii) Formation of resource groups on girls' education at block, district, and state level;

(iii) Provision of separate toilets for girls;

(iv) Training of teachers on gender aspects;

(v) Introduction of Village Education Committee (VEC) and Parent Teacher Association (PTA) in schools;

(vi) Preparation of textbooks free from gender bias;

(vii) Provision of funds for the construction of 40 seated girls' hostel in 396 Kanyashrams in the KBK districts;

(viii) Introduction of the National Programme for Education of Girls at Elementary Level (NPEGEL) in 165 educationally backward block of Orissa;

(ix) Opening of 2,875 model cluster schools for girls education under Sarva Sikshya Abhiyan (SSA)l; and

(x) Opening up of residential girls' school under Kasturba Gandhi Vidyalaya etc.

Gender Disparity Among Social Groups

There is a wide gender disparity in literacy rate among the social groups including Schedule Castes, Schedule Tribes and

General categories. Amongst different social groups the literacy rate in Orissa is the lowest in the case of Scheduled Tribes and the highest among general castes, as can be seen in theTable 12.1. Not only the gender disparity among the social groups but also there is a wide variation within the groups. The gender disparity in literacy is found to be the highest among the Scheduled Tribes and lowest in the case of general castes. Though there is improvement in the literacy rate since last thirty years still there is variation among the social groups. Within the Schedule Tribes the gender difference is highest and not satisfactory. All these are shown in the Table 12.1.

The present system of education in tribal areas of Orissa is unsatisfactory due to several factors. These can be categorized as: teacher absenteeism, the growing gap between teachers and the taught, apathy of teachers, unsuitable school timing, lack of participation of parents in the management of schools and the prevalence of physical punishment for students. Other factors of deterioration of the education system in tribal areas include: the growing inferiority complex of tribal students, increasing contempt for manual labour, different languages used at home and at school, apathetic attitude of non tribal teachers towards tribal students, their language and culture, lack of teacher training for addressing the bilingual classroom, lack of resources and academic support at the district and sub district levels, reluctance of tribal parents to send girls to schools, and the individualism of teachers.

Level of Educational Attainment and Gender Disparity

Though literacy is the most standard and popular indicator of development in a society, it does not by itself say anything about the level of educational attainment. In Orissa a significantly higher proportion of females is illiterate as compared to males as shown in Table 12.2. In addition, the proportion of males completing different levels of school education (primary, middle, and secondary) is higher than females, though the difference between these two narrows down with time. Still the illiterate rate among the female is 71.17 per cent in 1991 census. Not only this the other categories of educational level for female in comparison with male is lower. So all these indicators are shown there is a strong gender disparity in Orissa.

Table 12.1: Social Group and Sex-wise Literacy Rates (in per cent) in Orissa

Caste	*1971*				*1981*				*1991*			
	M	*F*	*T*	*GD*	*M*	*F*	*T*	*GD*	*M*	*F*	*T*	*GD*
SC	25.98	5.17	15.61	4.03	35.26	9.40	22.41	2.75	43.03	17.03	30.19	1.53
ST	16.38	2.58	9.46	5.35	23.37	4.76	13.96	3.89	27.93	8.29	18.10	2.37
General	49.35	20.37	35.02	1.42	58.15	29.84	46.03	0.95	63.50	39.54	51.77	0.61
Total	**38.30**	**13.92**	**26.18**	**1.75**	**47.09**	**21.12**	**35.37**	**1.23**	**52.41**	**28.83**	**40.80**	**0.82**

Sources: Government of Orissa, (2004) Human Development Report 2004, Orissa

Notes: Literacy rates have been calculated on the basis of literates and total population.

Table: 12.2: Distribution of Population by Educational Level in Orissa (in per cent)

Education level	*1971*			*1981*			*1991*		
	M	*F*	*T*	*M*	*F*	*T*	*M*	*F*	*T*
1	2	3	4	5	6	*7*	*8*	*9*	*10*
Illiterate	61.71	86.08	73.82	52.90	78.88	65.77	47.59	71.17	59.20
Literate without education	6.54	2.80	4.68	19.01	9.49	14.29	18.12	10.91	14.57
Primary	21.40	8.83	15.16	12.53	6.52	9.55	11.03	7.15	9.12
Middle	7.25	1.88	4.58	9.26	3.71	6.55	13.34	7.17	10.30
Matriculate	2.43	0.33	1.38	3.59	0.91	2.26	4.76	1.99	3.39
Higher secondary	–	–	–	0.99	0.21	0.60	1.83	0.70	1.27
Non technical diploma	0.11	0.01	0.06	–	–	–	0.13	0.03	0.08
Technical diploma	0.07	–	0.04	0.45	0.06	0.27	0.48	0.12	0.31
Graduate and above	0.49	0.07	0.28	1.27	0.22	0.75	2.72	0.76	1.76
All levels	100	100	100	100	100	100	100	100	100

Sources: Government of Orissa, (2004) Human Development Report 2004, Orissa.

Gender Disparity in Enrolment

Not only there is gender disparity in literacy and educational level but also there is a strong gender disparity in enrolment in the educational institute. Between 1950-51 and 2000-01, the enrolment of girls has increased at an annual rate of 7 per cent at the primary level, compared to 4.9 per cent for boys. The corresponding increase in growth rates for girls and boys at the upper primary level were above 10 per cent and 5.9 per cent, respectively. The gender parity index for enrolment in primary schools increased from 0.65 in 1980-81 to 0.71 in 1990-91, remained at the same level of 0.71 in 2000-01 and then increased to 0.90 in 2003-04. Thus, there has been a significant improvement in the gender parity index, although girls still lag behind boys in terms of enrolment ratios. Gender Parity Index in enrolment in elementary education in Orissa are shown in detail in the following Table-12.3.

Table 12.3: Gender Parity Index* in Enrolment in Elementary Education in Orissa

Year	*Primary*	*Upper Primary*
1950-51	0.26	0.80
1960-61	0.44	0.13
1970-71	0.53	0.32
1980-81	0.65	0.48
1990-91	0.71	0.62
2000-01	0.71	0.77
2002-03	0.87	0.80
2003-04	0.90	0.82

Sources: Government of Orissa, (2004) Human Development Report 2004, Orissa.

Note: *Ratio of number of girls to number of boys enrolled in school.

Gender Disparity in Gross Enrolment Ratio

The absolute figure on enrolment and their high growth rates do not say much about the progress, it is important to look at the percentage of children of school going age group who are actually attending schools. At the time of independency in 1947-48, the gross enrolment ratio in primary education in the state was 14 per cent whereas it is 28 per cent for boys and 1 per cent for girls. The Gender Parity Index was 0.04. Since then it is increasing but not satisfactory. By 1999-2000, this ratio increased to 108.8 per cent compared to 105.4 per cent in 1992-93. A sex wise increase in gross enrolment ratio in primary education was also found in different years, but the ratio was more for boys compared to girls. In the case of upper primary education, the gross enrolment ratio increased from 3.3 per cent in 1947-48 to 53.3 per cent in 1999-2000. The boys as had a more favourable ratio as compared to the girls. All these information are shown in detail in the following Table 12.4.

Health and Gender Disparity

Women are not only discriminated on the education front but also in terms of their access to health care facilities, reproductive rights, proper diagnosis of diseases, and nutritious food intake. Some of the studies found that the gender disparity in a place can well be judged through utilization of health services which are influenced by a range of factors, including physical proximity and cost. Not only this, the quality of care provided and the organization and delivery of health services can provide sufficient evidence for understanding the gender disparity in a place (Kumar, 2002; Sahay, 2004; Saha and Ravindran, 2002). Some of the empirical studies found that the utilization of health services is impaired by gender and social inequalities. There are specific physical or financial barriers in access to health services that arise because of one's gender or social status. It is not only the disparity between the male and female in the society but also the disparity can be well judged among the social groups as well as within the groups. Men and women members of different social groups experience any form of overt or covert discrimination within the health services, whether from service providers or because of the way services are organized. The organization off health services and allocation of resources accords a lower priority to women as compared to men, the poor and marginalized as compared to better of groups in society. Though this is the general scenario in Indian society, Orissa is not out of this scenario. According to NFHS-2 (1998-99), almost half (48 per cent) of the women population in

Table 12.4: Gross Enrolment Ratio in Primary and Upper Primary Schools in Orissa

Year	*Primary (Age group: 6-11 years)*				*Upper Primary (Age group: 11-14 years)*			
	Boys	*Girls*	*Total*	*Gender Parity Index**	*Boys*	*Girls*	*Total*	*Gender Parity Index**
1947-48	28.0	1.0	14.0	0.04	6.0	0.4	3.3	0.07
1950-51	28.0	7.0	17.0	0.25	7.0	0.5	4.0	0.07
1960-61	89.0	39.0	64.0	0.44	16.0	2.0	9.0	0.13
1973-74	93.6	56.5	75.4	0.60	31.0	11.2	21.8	0.36
1979-80	97.8	67.0	82.8	0.69	39.6	19.1	29.5	0.48
1986-87	109.1	82.8	96.2	0.76	56.9	26.7	40.7	0.47
1992-93	120.7	89.2	105.4	0.74	75.7	44.1	60.2	0.58
1998-99	109.5	79.8	94.9	0.73	64.8	37.4	51.3	0.58
1999-00	125.7	91.5	108.8	0.73	66.6	43.8	55.3	0.66

Sources: Government of Orissa, (2004) Human Development Report 2004, Orissa.

Note: *Ratio of number of girls to number of boys enrolled in schools.

Orissa suffers from nutritional deficiency, with a body mass index (BMI) less than 18.5. This problem is particularly serious for younger women, illiterate women, and women belonging to the SC and ST communities. Similarly, the prevalence of anaemia is very high among women in the 15-49 year age group and in children below 3 years of age. As a result, they are more vulnerable to diseases and sickness unlike their male counterparts. Their problems get aggravated due to the gender bias in health care access and practices. Those health facilities are available in Orissa are also less in comparison to all India where the health facilities for women is only 15 per cent.

A small number of studies, including two national surveys by the NCAER (1992) and the NSSO (1998) found that there is a gender difference in the utilization of health care services. It can be studied by examining the utilization of health care services for inpatient as well as out patient care. Not only this it can be examined by the proportion of illness episodes that are treated or the proportion of hospital beds occupied by women as compared to men. In Orissa there are a lot of instances for the utilization of health care services between men and women especially in the household level. Some of the statistics shows that there is a strong gender disparity in the utilization of health care services in Orissa.

From another angle the gender disparity in health can be studied by looking into the health expenditure pattern. A few studies are shown that more expenditure is spent per illness episode in men as compared to women. If we look into the age and gender wise expenditure in Orissa, it is found that there is a difference between men and women in different age group.

The vast majority of the studies that had any information on women's utilization of health services dealt with reproductive health services, especially pregnancy and delivery care and family planning. The overwhelming evidence in this regard relates to women's non-utilization of health facilities for delivery and low utilization of antenatal care. There is a clear gender bias against women in the acceptance of all modes of family planning and specifically with respect to sterilization in Orissa. A total of 1,345,586 sterilization cases were reported in Orissa during the period 1990-91 and 1999-2000. Of this, 12,98,873 (96.53 per cent) were tubectomy cases and only 46,713 cases (3.47 per cent) were vasectomy cases. Though vasectomy is a comparatively easier family planning acceptance technique, the percentage share of vasectomy cases has registered a marked fall, while the tubectomy cases, covering sterilization of women, have increased phenomenally. These are some of the instances of gender disparity in health status in Orissa.

Conclusion

There are a lot of measures taken for reducing the gender gaps since independence. Still there is lot of disparity not only in education and health but also in each and every field there is some type of gender disparity shown in India as well as in Orissa. There are some inter state differences in gender disparities, since it is varying with different levels and magnitude across India. In absolute term in India the number of male and female are more or less equal, so there should not be any gender disparity in any form. For the development of a country there should be equal opportunities for both the sex. There shouldn't be any kind of discrimination between sexes. As United Nations Secretary General Kofi Annan has stated 'Gender equality is more than a good in itself. It is a precondition for meeting the challenge of reducing poverty, promoting sustainable development and building good governance'. This recognition is currently missing in the state of Orissa. Transforming the prevailing social discrimination against women must become the top priority and must happen concurrently with increased direct action to rapidly improve the social and economic status of women.

Following are some of the policy measures proposed for consideration for reducing the gender disparity in the state of Orissa.

Policy Implications

Since gender inequalities harm people's well being and constrain a country's development prospects, there is a compelling case for public action for promoting gender equality. The state has a crucial role in improving the well being of both women and men. The state can obtain substantial social benefits by improving the absolute and relative status of women and girls. Public action is particularly needed to change social, political, economic, and legal institutions that perpetuate gender inequalities. Some of the measures that would help in promoting gender equality are outlined below.

General Policy Measures

(i) Institutions may be reformed to establish equal rights in terms of legal, social and economic opportunities for women and men.

(ii) Economic development can be expedited to strengthen incentives for equal resources and participation.

(iii) Active policy measures can be taken to reduce persistent disparities in control over resources and political voice.

Specific Measures

The combined effects of institutional reforms and economic development usually take time to be realized. Hence, in the short and medium term, active measures are required. These are as follows:

(i) Reducing the cost of schooling;

(ii) Redesigning financial institutions by simplifying banking procedures and delivering financial services closer to homes, markets and workplace;

(iii) Implementing gender sensitive land reforms that provide joint ownership to both husband and wife, or enable women to hold independent land titles;

(iv) Providing public support for out of home child care services.

Most of the above measures will reduce the amount of time spent by women and girls on their unpaid work. Thus, it will allow women to engage themselves in more productive and paid work and train girls to attend schools. There is strong case for state intervention to promote gender equality through these measures.

Other measures more specifically relevant to Orissa are:

(i) Raising social awareness about the importance of female education.

(ii) Spread of female education through scholarships and other incentives, increasing investment on female education and installing separate infrastructure like ladies toilets and common rooms in the existing educational institutions.

(iii) Encouraging vasectomy as a family planning instrument to reduce the total burden of family planning borne by women.

(iv) Ensuring adequate supply of fuel, food and water at the village level and setting up of baby care and child care centers.

(v) Proper implementation of Equal Remuneration Act for equal wages and organisation of women wage workers.

(vi) Proper care of single, poor and elderly women by the state.

REFERENCES

Government of India (2001) Census of India: Provisional Population Totals, Series-22: Orissa, Directorate of Census Operations, Orissa.

Government of Orissa, (2004) Human Development Report 2004, Orissa. Planning and Co-ordination Department, Government of Orissa.

International Institute for Population Sciences (IIPS) and ORC Macro (2001), National Family Health Survey (NFHS-2), India, 1998-99: Orissa, IIPS, Mumbai.

Kateja, Alpana (2001) Gender Disparities and Human Development: A Regional Study. Indian Journal of Regional Science, Vol. XXXIII, No. 2, pp. 43-52.

Kumar, Rachel (2002) Gender in Reproductive and Child Health Policy, Economic and Political Weekly, Vol. XXXVII, No. 32, pp. 3369-77.

Novack, David R. and Lesley Lazin Novack (1997) Gender Disparity in Advanced Industrial Society: Female Male Relations at a Recently Co-educated College. Human Organization, Vol. 56, No. 2, pp. 245-52.

Pati, Sanghamitra (2003) Health and Human Rights Inextricable Link. Social Welfare, Vol. 50, No. 4, pp. 15-19.

Prasad, R. Sivaram and Talluru Srinivas (2003) Women and Health Issues. Social Welfare, Vol. 50, No. 4, pp. 13-14.

Saha, Shelley and Sundari Ravindran (2002) Gender Gaps in Research on Health Services in India. Journal of Health Management, Vol. 4, pp. 185-214.

Sahay, Aparna (2004) Gender Disparity and Women's Autonomy. Prashasanika, Vol. XXXI, No. 1, pp. 85-92.

Soundari, M. Hilaria and M.A. Sudhir (2003) Gender Disparity in Tamil Nadu. Social Welfare, Vol. 49, No. 12, pp. 24-27.

Sudharani, K. et al. (2003) Awareness of Women About Their Rights. Social Welfare, Vol. 50, No. 4, pp. 20-22.

United Nations Development Programme (UNDP) (2005) Human Development Reports 2005. Oxford University Press, New York.

World Bank (2000) World Bank Policy and Research Bulletin, July – September, Vol. II, No. 3.

13

The Status of Women in Orissa

Mr. Santosh Kumar Sahu

Introduction

The population of the state as per Census of 2001 is 3.67 crore which forms about 3.58 per cent of the total population of the country. The female population is about 49.29 per cent in the state. The decadal growth rate of population between 1991 and 2001 is 15.94, which is one of the lowest among the states of India. Out of the total population the scheduled castes form about 16.20 per cent and the scheduled tribes of 24.21 per cent. The state is primarily a rural state where about 87 per cent of people reside in the rural area and the urbanization forms only about 13 per cent, one of the lowest percentages of urban population among the states (Census, 2001).

The sex ratio of the state is 972 according to Census of 2001 which gradually slowed down from that of 1086 from the year 1921. It can be verified from Table-13.1 that the sex ratio of Orissa remained higher than that of the country throughout the twentieth century. Up to the year 1921 the sex ratio remained above 1000 in the state, but thereafter it has been declining. It is found that the difference between the sex ratio of Orissa and India is gradually narrowing down during the last century, from 131 in 1921 to 39 in 2001.

Rural and Urban Sex Ratio

The declining trend is observed in case of sex ratio of rural population of Orissa, but interestingly the sex ratio of urban Orissa showed an increasing trend since 1961. In the year 1961 onwards till 1991, the sex ratio in urban Orissa has increased from 807 to 866.

Table 13.1: A Comparative Picture of Sex Ratio of Orissa and India

Year	*Orissa*	*India*	*The Gap*
1921	1086	955	131
1931	1067	950	117
1941	1053	945	108
1951	1022	946	76
1961	1001	941	60
1971	988	930	58
1981	981	934	47
1991	971	927	44
2001	972	933	39

Source: Census of India, 2001.

On the whole, throughout the twentieth century the sex ratio in urban Orissa remained less than in rural Orissa. The reason behind this contradiction may be due to the rapid growth of urbanization and industrialization in the state since 1961. Another trend reveals that the rural sex ratio up to 1971 remained more than one thousand and thereafter remained below 1000. On the other hand, in urban population from the beginning of 1921 it remained less than 1000. Apart from the declining sex ratio, Orissa shows a stagnant Crude Birth Rate (CBR) of 26.2 per 1000 population. In India the urban sex ratio shows a declining trend in comparison with rural sex ratio. Orissa's CBR is lower than the national average of 27.4 but it is still higher as compared to many states like Kerala (17.8), Punjab (23.5), Maharastra (23.2) and Tamilnadu (19.2).

Table 13.2 provides separate figures for sex ratio of rural and urban population.

Table 13.2 : Trends of Rural and Urban Sex Ratio in Orissa

Year	*Rural*	*Urban*	*Total*
1921	1089	963	1086
1931	1071	924	1067
1941	1058	914	1053
1951	1029	851	1022
1961	1015	807	1001
1971	1002	845	988
1981	999	859	981
1991	988	866	971
2001	986	895	972

Source: Census of India, 2001.

Infant Mortality Ratio (IMR)

Death rate in Orissa is one of the highest in India with 11 per 1000 population in comparison to national average of 8.9. This is also higher than all other states. The age specific death rate shows that mortality of girls occur somewhere between the growing age and period of adolescent, which is regarded as most important preparatory period of girls life for motherhood. The IMR has been taken as one of the important measurement of health status of population and indicator of women's development in general. The less the number of infants die in a year, higher in the status of women.

The IMR is sharply declined in Orissa since 1991, but the ratio is highest in Orissa among the states with 96 deaths per 1000 population. In 1997 it is higher about 35.21 per cent to the national death rate of 71. Up to 1995 the IMR in Orissa remained three digits with 103. But to-day the death rate still remains high with 100 in the rural areas of the state.

Life expectancy at birth is another measurement, which indicates the general health of a population. The life expectancy at birth of females in Orissa remained at 58 years during the period of 1991-96 as against 60 years for the males during the same period. The all India figures of life expectancy at birth show a different trend; females live longer than males, in the former case it is 61 years while the latter live up to 60 years. According to the Census of India 1991, the mean age of women in Orissa at the time marriage was 19.1 years.

Literacy

According to the Census of 2001, the female literacy in Orissa is 50.97 per cent as against literacy of about 75.95 per cent among males. The combined general literacy of male and female together is 63.61 per cent in 2001.

There is wide gap between male and female literacy in the state. Since independence the percentage of female literacy increased from 4.52 to 50.97 per cent over five decades. In comparison to this the male literacy in the state increased from 27.32 to 75.95 per cent in 50 years.

There is a wide difference of literacy rate among females in the different districts of Orissa. The district of Khordha registered highest per cent of female literacy of 71.06 per cent in comparison to the tribal district of Malkangiri with a lowest per cent of literacy rate of 21.02 per cent. The four tribal districts of Koraput, Rayagada, Malkangiri and Nabarangpur have 24.81, 24.31, 21.28 and 21.02 per cent of literacy rate respectively.

The growth of male and female literacy in Orissa from the year 1951 to 2001 is presented in Table 13.3.

Table 13.3: Growth of Male and Female Literacy in Orissa 1951-2001

Year	*Persons*	*Male*	*Female*	*Difference*
1951	15.80	11.03	4.50	6.08
1961	21.66	34.68	8.65	26.03
1971	26.18	38.29	13.92	24.37
1981	34.23	47.10	21.92	25.98
1991	49.00	63.09	34.68	28.41
2001	63.61	75.95	50.97	24.98

Source: Census of India 2001.

Literacy and Poverty

The district figures of female literacy and the people below the poverty line in the districts show that there is high positive co-relation between the two. Districts with low literacy among the women have high incidence of poverty. The district of Koraput with 24.41 per cent of female literacy has high incidence of poverty with 83.81 per cent of people below the poverty line. On the other hand, Ganjam with high rate of female literacy of 47.70 has 55 per cent of people below the poverty line.

Percentages of female literacy and percentage of people below the poverty line in the 15 districts of Orissa is presented in Table 13.4.

Table 13.4: Female literacy and Poverty in the Districts Figures in Per cent

No.	*District*	*Female Literacy 2001*	*People below Poverty 1997*
1.	Nabarangpur	21.02	73.66
2.	Malkangiri	21.28	81.88
3.	Rayagada	24.30	7.02
4.	Koraput	24.81	83.81
5.	Nuapada	26.01	85.70
6.	Gajapati	28.91	61.38
7.	Kalahandi	29.56	62.71
8.	Kandhamal	36.19	78.12
9.	Mayurbhanj	38.29	78.42
10.	Balangir	39.27	61.06
11.	Boudh	39.78	80.20
12.	Kendijhar	46.71	76.96
13.	Sonepur	47.28	73.02
14.	Deoghar	47.56	78.79
15.	Ganjam	47.70	55.00

Source: DSH, Orissa 1999.

Education

Education is universally accepted as an instrument of improvement in the status of women. It is also considered as stepladder for occupational and social mobility. Post independence period has witnessed a significant improvement in women's education in the state though the overall situation still remains far from satisfactory.

Enrolment of girls in primary schools marginally declined from 1990 to 1996 while comparing the four stages of Indian educational system that is primary, middle, secondary and college level. Enrolment of girls in the primary schools remained as low as nearly·17 per cent in comparison to the enrolment of boys throughout these years. The enrolment of girls during these years increased about 5 per cent in middle school level, 4 per cent in the high school level and three per cent higher education.

In the middle level education, girls-boys ratio in enrolment is nearly 1:2. It was also same for the secondary level enrolment of both the sexes. In the higher education the gap of enrolment in boys and girls is still higher. The enrolment of girls is as fewer as one third of boys; comparing from primary to higher education the number of girls has been decreased consistently from 41.36 to 25 per cent years during the period of 1991 to 1996.

Enrolment of boy and girl students in the educational institutions of the state from the primary school level to the college level education in terms of per cent is presented in Table 13.5.

Table 13.5: Women's Education in Orissa 1991-96 Figures in per cent

Year	*Primary School*		*Secondary School*		*Middle School*		*Higher School*	
	Boy	*Girl*	*Boy*	*Girl*	*Boy*	*Girl*	*Boy*	*Girl*
1990-91	58.46	41.53	66.62	33.24	61.86	31.88	76.49	23.13
1991-92	58.45	41.54	66.45	33.54	61.91	38.08	76.36	23.27
1992-93	58.44	41.55	65.32	34.67	62.00	37.99	75.17	24.82
1993-94	58.44	41.45	64.91	35.08	62.02	37.97	70.91	28.78
1994-95	58.63	41.36	63.72	36.15	62.06	37.93	75.00	25.00
1995-96	58.63	41.36	63.38	36.61	61.96	38.03	74.77	25.00

Source: Statistical Abstract of Orissa, 1996-97.

Health Status

Eight Indicators can be taken for the measurement of the degree of health status of the population. The overall measures of health status for Orissa shows that it is far from satisfactory. These eight indicators for Orissa are presented Table-13.6.

Table 13.6: Indicators of Health Status of Orissa

No.	*Indicator*	*Figure*
1.	Sex Ratio (2001)	972
2.	Birth Rate (1996)	26.8
3.	Death Rate (1996)	10.7
4.	Female Literacy (2001)	50.97
5.	Infant Mortality Rate (1998) (Per 1000 live births)	96
6.	Total Fertility Rate 91993) (per Women)	3.1
7.	Mean Age at Marriage (1991)	19.1
8.	Female Life Expectancy (1996)	58

Source: Economic Survey of Orissa, 1997-98 and Census of India 2001.

Based on a weight-for-height index, about 48 per cent of women in Orissa are under-nourished. Nutritional deficits are particularly serious for women in rural areas and in disadvantage socio-economic groups. Women who are undernourished themselves are also much more likely than other women who have children who are undernourished. Anaemia is a serious problem among women in every population group, with prevalence rates ranging from 44 to 75 per cent. On an average, about 65 per cent of women in Orissa have some degree of Anaemia.

Nutrition of Women

The nutritional status of women has not improved substantially and in some regions has deteriorated over the past few years faster than the nutrition of men. The per capita calorie intake in the country as a whole declined from 2445 in 1961 to 2179 in 1971. The worst sufferers are the women as agricultural labourers, living in slums and living in the remote tribal areas. Malnutrition among women has further been aggravated due to repeated pregnancy and lactation. During pregnancy a woman requires about 2,500 calories of nutritious food but receives only about 1,440. Three main nutrients required for pregnant women are protein, iron and calcium. It is found that the pregnant women are not getting their required nutrients; the efficiencies are as high as 80 per cent in case of calcium followed by iron and proteins 55 and 33 per cent respectively.

The nutrients needed and provided to the pregnant women and the deficiencies thereof are presented in Table 13.7.

Table 13.7: Nutrients Requirement for Pregnant Women Figures in Grams

Food	*Needed*	*Provided*	*Deficiency Per cent*
Protein	55	37	32.72
Iron	40	18	55.00
Calcium	1	0.2	80.00

Source: The ICMR Report 1981.

Family Planning

The rate of fertility is gradually in the declining trend in Orissa. Throughout their child-bearing period the fertility rate remained 2.9 in the year 1992-93 (NFHS-1) currently reduced to 2.5. Efforts to encourage the trend towards lower fertility rate have to concentrate in those specific groups in the population whose fertility rate is higher than the state average.

It has been identified that in the following categories of women the rate of fertility is higher than other women. These categories are: rural women, illiterate women, poor women along with women of specific communities schedule castes and schedule tribes.

Fertility is more among the young women. It is found that about 16 per cent of the total fertility is from the females in the age group of 15-19 years. Studies confirm that health and mortality risks increase when women give birth at these young ages, both for the women themselves and for their children.

Modern women comparatively are more health conscious and prefer smaller families. About 5 per cent of the mothers reported that (NFHS-2) they preferred not to have child within three years of their marriage, while about 12 per cent of mothers expressed that they would have preferred to delay the pregnancy. About 35 per cent of mothers who had there children and another 31 per cent of mothers with children four and more considered two-child norm as ideal for a family.

In 1993-94 about 36 per cent of women were using contraceptives but now the use increased to about 47 per cent. There is also an urban-rural difference in use of contraceptives. About 54 per cent of urban families use contraceptives in one form or other but in the rural families it is lower to about 46 per cent.

Adoption of contraceptives increases with the increase of age of the women and also number of children in the family. About 9 per cent of the newly married women adopted contraceptives of temporary nature to delay the first birth of child. The use of contraceptives by the women from Orissa is highest of 69 per cent in the age group of 40-44 years.

About 9 per cent of the families with two children but without a son adopted male or female sterilization, while on the other hand, about 18 per cent of the families adopted permanent sterilization method to restrict their family.

The Government of India has been consistent in spreading the message of family planning measures from 1960 onwards. But now according to the national Family Health Survey, the family planning message reached the specific groups as presented in Table-13.8. It can be verified that out of six categories of under privileged groups the reaching of message ranged from 28 to 63 per cent. The lowest percentage of social group women with 28 per cent is the scheduled tribe women. Other than these groups, that constitute mainly higher income group of people and the people other than scheduled and backward population, the message of family planning reached to the maximum of about 78 per cent.

Table 13.8: Reception of Family Planning Message

No.	*Social Group*	*Per cent*
1.	Illiterate women	41
2.	Lower Income Women	42
3.	Schedule Caste Women	56
4.	Schedule Tribe Women	28
5.	Backward Class Women	63
6.	Other Groups	78

Source: National Family Health Survey-II, 1999.

Work Participation

Due to some historical and social reasons women have lagged behind men relating to their productive participation in the society. They have mainly concentrated in the role of reproductive and service sectors whereas men have participated in productive activities and become the bread earners in the family. It has given men enough power to dictate over women both inside and outside the family. After fifty years of independence, even though there are no constitutional restrictions for women to participate in the remunerative jobs alike men, but due to some traditional social reasons they still remain inside the house doing only household chores instead of participating in the process of economic development.

As per 1991 census, about 20 per cent of females were working population in Orissa as against 80 per cent in case of males; within these 20 per cent of female workers again 59 per cent are categorized as main workers and the rest of 41 per cent as marginal workers.

In the agricultural sector all the agricultural operations such as weeding, transplantation, threshing, harvesting of different crops, processing of paddy etc. are proximately done by women marginal workers. According to 1991 census, about 80.9 per cent of female workers were engaged in agriculture workers. Men do ploughing of land and other strenuous work. But in this sector also, about 25.8 per cent of female workers were also engaged in agriculture as marginal workers.

Employment of women in the organized industrial sector in both private and public enterprises is very much limited. It may be observed from Table-13.9 that in both the sectors the participation of women workers is on increasing trend. In the year 1998 the total women workers engaged in the organized sector of Orissa remained about 11.49 per cent, which does not reflect women's entry into the organized sector on a large scale.

Table 13.9: Employment of Women in the Organized Sector of Orissa Figure in Per cent

Year	*Public Sector*	*Private Sector*	*Total*
1990	8.23	12.04	8.77
1991	8.74	12.24	9.20
1992	8.78	12.73	9.30
1993	9.71	10.98	9.85
1994	10.03	12.42	10.32
1995	10.37	12.04	10.63
1996	9.43	13.40	9.87
1997	11.07	11.31	11.11
1998	11.54	11.12	11.49

Source: Economic Survey, Orissa 1998-99.

The women workers are concentrating more in the private sector than in the public sector enterprises. It speaks availability of women as cheap labour to private employers. Women face more vulnerability at work place in private sector than in public sector employment. However, it has been found that women's participation in public sector enterprises is gradually increasing from 8.23 per cent in 1990 to 11.54 per cent by the year 1998. On the other hand, the employment of women is in the declining trend from 12.4 per cent to 11.12 per cent. In the terminal year of 1998 the participation of women in both the sectors almost remained same. This low level of employment of women in the organized sector of Orissa is mainly due to low level of educational status of women in the state. But the growth of female education and coupled with social rethinking on gender relations may result in fundamental change in the pattern of power relations within families and outer world (Baral & Das, 1999).

The socio-economic indicators show that there is an increase in female literacy, and work participation in the private sector. These indicators reveal that women are better placed in the society. More and more girls are attending school even in the rural areas. Quite often girls are married at a tender age, which may be one of the reasons for the dropout in education while for the dropout poverty plays great role too. There is still discrimination between a boy child and a girl child. Sharing of housework is still a taboo among boys. Cases of murder, physical and mental torture of women are increasing day by day. Even in well-to-do families the status of women has improved but little. Women are harassed in the work place both in organized and unorganized sectors. Therefore, keeping sustainable development as the view point of growth women are to be more educated and should enjoy more and more facilities that they can contribute their full effort in constituting a developed and humanitarian society.

REFERENCES

Chakravarti, B.K., Slum Problems in India, Genesis and Possible Solution, in Roy, P. & S.D. Gupta (Ed.), Urbanization and Slums, Hur Anand Publications, New Delhi.

Flower, John M. (1970), Rediant Living, Original Watchmen Publication House, Poona.

Mahta Ashoke, (1973), Indian Today, S.Chand & Co., New Delhi, p. 65.

Panda P.K. (1997), Female Head Ship, Poverty and Child Welfare, A Study of Rural Orissa, EPW, Oct 25,1997, p. 81.

Pillai Gita, (1995) ICCW Journal, New Delhi, June-Dec, Vol. III, No-1 & 2, p. 25.

Primary Healthcare: Report on the International Conference on the Primary Healthcare, Health for All, S 1.1, Alma Ata, USSR.

Sen Amaretya, Women, Tech. and Sexual Division of Labour, UNCARD Review, Vol. 6.

Silva Kumar, A.K., Health as Development, EPW, April-17, 1993, p. 772.

WHO, Report of the WHO Meeting in Women and Health, Publes, Edinburgh, May 25-27, 1985.

Orissa Census Report, 1971-2001.

India Census Report, 1961-2001.

14

The Way They Are

Exploring the Sexual Behaviour of Urban Youth in Puri, Orissa

Mr. Manas Ranjan Pradhan and Dr. Usha Ram

Introduction and Context

Youth (15-24 years) who largely define the socio-economic and political future of a population are 20 per cent of the Indian population (RGI, 2001 and UN, 2001). Although the present cohort is more urbanised and better educated, social vulnerabilities persist and transitions to adulthood are frequently marked by premature exit from school, entry into the labour force and strongly held gender norms (Jejeebhoy and Sebastian, 2004). Further, young peoples' vulnerability caused by their young age, coupled with lack of and/or poor knowledge on matters related to sexuality, reproductive health, and their inability or unwillingness to use family planning and health services puts them at a significant risk of experiencing negative consequences (Jejeebhoy and Sebastian, 2003; Mamdani, 1999 and Singh, 1997).

Internationally, young peoples' sexual health is a major concern both because of an urgent need to reduce the high level of unwanted pregnancy and sexually transmitted diseases and because of a desire to improve less tangible aspects of health such as psychological well being. Young age coupled with uncertainty and emotional insecurity is likely to lead them for sexual experimentation. Greater reliance on inappropriate information sources or the avoidance of seeking professional guidance adds to the differences further. In the process of discovering their sexuality, youth often experiment and are thus more vulnerable to risky sexual acts. Moreover, cultures where talking about sex with young people continues to be a stigma, may further contribute to the risk taking behaviour (Miller and Whitaker, 2001).

In South Asian societies, the general sexuality norms that govern adolescent boys and girls are almost diametrically opposite. Gender socialization operates in a manner that promotes male sexuality while constraining female sexuality through dominant social institutions and ideological practices (Abraham, 2002). Further, in a country like India where talking about sex continues to be a taboo and gender differences prevail in almost all aspects of life, examining interlink of reproductive and sexual behaviour especially of the youth holds an important and crucial place. Youth, who are less addressed, demand special attention not only of the policy makers but also of the researchers interested in the field. Every third person in the age group 15-24 years in India is married, widowed, divorced or separated (RGI, 2001). It has been found that there is a strong relationship between socio-demographic problems like infertility, unwanted pregnancy, sexual abuse and sexually transmitted infections including HIV/AIDS, with youth sexual behaviour (Varga, 1999 and Radhakrishna et al., 1997). Again, approximately 16000 people world-wide are infected with HIV/AIDS each day and majority of all new infections occurs among those aged 15-24 years (Advocates for Youth, 1999). Further, studies have also reported that young people form a significant segment of those acquiring sexually transmitted infections and those infected by HIV/AIDS (Verma et al., 2004; NACO, 1994 and Ramasubban, 1992).

While Draft National Youth Policy (GoI, 1997) has placed young men and women as a priority group and gender justice as major thrusts of the policy, the National Health Policy (GoI, 2002) highlights adolescent and youth health as a strategic focus in achieving socio-demographic goals. In addition, the Millennium Development Goal 2000 (UN, 2003) stresses on condom use at high-risk sex besides percentage of population aged 15-24 years with comprehensive and correct knowledge of HIV/AIDS as one of the indictor to attain its goals. Besides this, the International Conference on Population and Development, Cairo (ICPD, 1994) has affirmed the sexual and reproductive rights of the adolescents and the need to make them aware of these rights. It has also identified gender and sexuality as the two core subjects and factors of development. Again, empowering women and encouraging male involvement besides addressing the needs of disadvantaged and under served population group are some of the strategic themes delineated by National Population Policy of India (GoI, 2000).

In its wider sense, the present research attempts to understand various dynamics of the sexual behaviour of male youth in the city of Puri (Orissa). The specific objectives of the present analysis are to explore the cultural and contextual factors related to premarital sexual behaviour and to understand the sexual health problems besides the treatment seeking behavior of the youth.

Data and Methods

Sample

The study is based on a community level data collected during later part of the 2003. At first author himself has interviewed all 156 male respondents aged 15-24 years. Besides a structured interview schedule, various qualitative techniques like social mapping, free listing and in-depth interviews have also been used for collecting relevant data. A multi-stage sampling design has been adopted for selecting the respondents for the personal interview. At the first stage of sampling, the city is divided into four regions on the basis of geographical location and the unique features inherited by the areas. At the second stage, two wards from each region are selected randomly. At the third stage, one Census Enumeration Block (CEB) from each ward is selected randomly. Thus a total of eight CEBs from eight wards are selected for the fieldwork. After selection of CEB, 25 households (having at least one male member in the age group 15-24 years) from each selected CEB are randomly selected ensuring representation of the CEB. In view of the non-response, over sampling of the sample size is done by 25 per cent. At the last stage of sampling, 20 respondents are selected (one each from the selected households). For the in-depth interviews, two respondents from each CEB exhibiting high-risk behaviour are selected purposively. It is worth mentioning that necessary ethical guidelines like informed consent of the respondents or their guardians (in case of minors) are taken into consideration during data collection.

Variables Used and Analysis

The dependant variable in the present analysis is premarital sex and is defined as 'sexual intercourse with a female before marriage'. Additional questions are asked to determine the contexts of sexual debut, number of sexual partners, sexual history of the partner(s), condom use and subsequent sexual acts. As the analysis aims to examine the association between cultural and contextual factors with premarital sex, a number of variables are taken for the analysis. Age is usually a strong predictor of sexuality and has found to be significant in earlier studies on sexuality (Abraham and Anilkumar, 1999 and Djamba, 1997). In the present study, we have divided the sample into two age groups i.e. 15-20 and 21-24 years. Education undoubtedly influences youth behaviour and thus included in the present analysis (Brewster et al., 1998). It is divided into 'up to high school' and 'above high school' for the multivariate analysis. Besides this, occupation of the respondents that is expected to have relation with premarital sex is divided into three categories like 'student', 'working' and 'not-working' and included in the analysis (Abraham and Anilkumar, 1999).

In a country like India, respondent's caste needs special attention and thus is included in the analysis. It is divided into 'scheduled caste (SC)/ scheduled tribe (ST)/ other backward caste (OBC)' and 'other general caste'. Type of family has already proved as an important variable influencing youth sexual behaviour and thus included in the present analysis as 'joint family' and 'nuclear family' (Miller et al., 2001). In addition to this, another important variable i.e. presence of sister in the family has been included in the analysis as it is expected to influence the sexual behaviour of the sibling. It is divided into two categories like 'have sister' and 'do not have sister'. Like this income of the family is considered to be an important factor and included in the analysis as 'Low', 'Medium' and 'High'. Similarly, to capture the intercity variation, region[1] as a variable has been used. Although the city is divided into four regions, in view of the sample size, it is merged to two

1. Region 1: Extreme north of the city adjacent to beach; Region 2: Around the temple; Region 3: Connecting beach from Baliapanda to Penthakata and exposed more to western culture; Region 4: located in the central part of the city.

regions viz: 'region1&3' and 'region 2&4' (keeping in mind the geographical locations as well as the common features inherited by them) for multivariate analysis. Again, though religion matters a lot in influencing sexual behaviour, in the present analysis it has not been included as most of the respondents (96 per cent) are Hindus (Hill et al., 2004).

Two composite indices namely 'standard of living' of the household and 'life style index' has been constructed as the important determinant of the sexual behaviour of the youth. The standard of living index (SLI) has been constructed by taking into account seven variables—'type of house', 'ownership of house', 'source of lighting', 'source of drinking water', 'type of toilet', 'type of fuel', and 'ownership of land'. The variables are given scores ranging between '0' and '4' according to the intensity in a five-point scale (lowest to highest) and then were summed up to get the total value of the index. After obtaining the composite index, it is divided into three groups of low, medium and high by using the formula: (maximum-minimum)/3.

Similarly, the life style index (LSI) has been constructed on the basis of the seven variables—frequency of 'smoking, drinking alcohol, chewing tobacco, visiting beer bars, watching blue films, visiting commercial sex workers (CSW) and taking Bhang (a local intoxicant)'. The variables are given scores ranging between '0' and '4' according to the intensity in a five-point scale (lowest to highest) and then were summed up to get the total value of the index. The respondents obtaining '0' were considered as no risk group. After that the composite index is divided into two sub-groups of those at 'moderate risk' and 'high risk' by the formula: (maximum-minimum)/2. Here it should be mentioned that during the construction of the index, the researcher has verified the composite index with and without including the variable 'frequency of visiting commercial sex worker' (keeping in view the possibility of visiting CSW without having any other risk indicators). But as the result does not show any major difference, the researcher has decided to retain the index based on all seven variables as mentioned earlier.

The quantitative data has been analyzed through 'SPSS' and 'Atlas ti' has been taken for analyzing the in-depth interviews. Our analytical approach includes both descriptive data as well as bivariate and multivariate analysis. The descriptive statistics shows the distribution of respondents by the important variables. The bivariate analyses examine the association between each variable and dependant variable i.e. premarital sex. Finally, logistic regression has been carried out to find out the extent that the independent variables have any role in predicting the premarital sex.

Findings and Discussion

Profile of the Respondents

The data in Table 14.1 show that nearly 15 per cent of the respondents in the study are minors (i.e. less than 18 years of age) and about 39 per cent of them are in the age range of 18-20 years (i.e. below the legal age at marriage for men in India). The mean age of the respondents is about 20.3 years. While more than 60 per cent of the respondents are literate up to higher secondary level, the remaining have completed more than 12 years of schooling. A little over one-third of the respondents are students and more than 50 per cent are working either in formal or informal sector. The table clearly shows that nearly 38 per cents of the youth covered in the present study are coming under 'high risky life style'. Most of the respondents are Hindus (96 per cent) and belong to general caste (59 per cent). Respondents are almost equally distributed among the four regions visited in the study. About 66 per cent of the respondents live in the nuclear families while 30 per cent live in the joint families. Remaining about 5 per cent live either alone or with friend(s). More than half of the respondents have at least one sister in the family. Further, most of the respondents are from medium income families (61 per cent) and households coming under high standard of living (58 per cent).

Table 14.1: Socio-demographic Profile of the Respondents

Socio-demographic characteristics	*Per cent (N=156)*
1	2
Age (in years)	
15-17	15.4
18-20	38.5
21-24	46.1
Education	
Up to middle*	14.1
High school to higher secondary	51.9
Above higher secondary	34.0
Occupation	
Student	35.3
Working	54.4
Not working	10.3
Life style index	
No risky	21.2
Moderately risky	41.0
High risky	37.8
Religion	
Hindu	96.2
Muslim/Christian	03.8
Caste	
SC/ST**/OBC	41.0
Others	59.0
Region	
Region 1	25.6
Region 2	23.2
Region 3	25.6
Region 4	25.6
Type of family	
Nuclear	66.0
Joint	29.5
Living alone/with friends	04.5
Presence of sister	
Have sister	44.2
Do not have sister	55.8

(Contd...)

1	2
Income of the family* **	
Low	28.8
Medium	60.9
High	10.3
Standard of living of the household	
Low	19.2
Medium	23.1
High	57.7
Total	**100.0**

*Includes two illiterate respondents **Includes one scheduled tribe respondent ***Low income signifies annual income up to Rs. 45,000/-, Medium income signifies income of Rs. 46,000-92,000/- and High income signifies income of more than Rs. 93,000/-.

Table 14.2: Premarital Sex by Selected Socio-demographic Characteristics

Socio-demographic characteristics	*Per cent (N=156)*
1	2
All	**35.9**
Age (in years)	
15-17	12.5
18-20	18.3
21-24	58.3
Education	
Up to middle*	72.7
High school to higher secondary	34.6
Above higher secondary	22.6
Occupation	
Student	10.9
Working	47.1
Not working	62.5
Life style index	
No risky	00.0
Moderately risky	31.3
High risky	61.0

(Contd...)

1	2
Caste	
SC/ST**/OBC	31.3
Others	39.1
Region	
Region 1	45.0
Region 2	38.8
Region 3	38.8
Region 4	25.0
Type of family	
Nuclear	29.1
Joint	45.7
Living alone/with friends	71.4
Presence of sister	
Have sister	18.8
Do not have sister	49.4
Income of the family	
Low	42.2
Medium	31.6
High	43.8
Standard of living of the household	
Low	43.3
Medium	38.9
High	32.2

*Includes two illiterate respondents **Includes one scheduled tribe respondent.

Premarital Sex: Contexts and Vulnerability

About 36 per cent of the youth have reported premarital sex (Table-14.2) with a mean age of about 20 years at the premarital sexual debut. Nearly 13 per cent of the minor respondents (below age 18 years) have also found to be sexually active. The corresponding figure was slightly higher (about 18 per cent) among those aged 18-20 years (major but still below the legal age at marriage). As one would expect the share of respondents who have made sexual debut prior to the survey date was much higher (58 per cent) among those aged 21 years or older. Further, premarital sex is found to be negatively associated with the literacy status of the youth. It is observed that more respondents who are literate up to middle reported to have the sexual experience. The data also suggest that around 11 per cent of the students reported premarital sex. Relatively higher proportion of the youth belonging to general caste and those maintaining a high risky life style has also reported premarital sex as compared to their respective counterparts.

The regional variation within the city comes out from the findings as 45 per cent of the respondents from region-1 reportedly have had premarital sex compared to 25 per cent of region-4. It may be noted that incidence of premarital sex is higher among respondents living alone or with friends (5 out of 7 respondents coming under this category are involved in premarital sex) and who do not have sister in the family. ......*'Once he (my friend) had arranged a programme in his rented house. I learnt about his plans, only after reaching there at around 9 O'clock late in the evening. He told me to have sex with the girl he has invited to his place as there is no harm in it and since nobody except him would know about it. Though I was not interested at first, he persuaded me. By the time, I became little bit excited about the whole thing. I wanted to experience the whole thing, which till that time I had seen in the movies. So I did it.............. there' (20 year old fisherman)'*. Further, relatively higher proportion of youth belonging to low income and high income families has reported premarital sex as against of those belonging to medium income families. It has also observed that the involvement is relatively more among the respondents from households of low standard of living.

Table 14.3 reveals that about one in every fifth cases, the girl friends/ fiancée is the partners at sexual debut. Qualitative findings focus on the fear of losing the partner as well as peer pressure sometimes leads youth to indulge in premarital sex. The likelihood of sexual relationships with Bhauja[2] is recurring theme in Indian folklore and also in sexual case histories collected in several other studies (AIMS, 1998 and Goparaju, 1994). The present research seems to fall in these lines as more than one-third of the youth have their premarital initiation with a relative or married female especially their neighbours' wives. To quote one 19 year old respondent, *'...that night after dinner she called me to her room as other members of her family were away. When I entered her room by*

2. Refers to wives of older men in the community particularly in the neighborhood.

pushing the door slowly, she was changing her clothes and I came back. After sometime she called me slowly and asked why I returned that time. I did not know that she was aware of my presence that time. Then in between talks she kissed me and touched my genitals. At first I was not ready but she undressed me and after that we did it. She assured me not to worry, as it was her safe period (according to her)'. The table also shows that marital status of every third partner of sexual debut is unknown.

Table 14.3: Details of Sexual Debut

Characteristics	*Per cent (N=61)*
Age of the partner	
Younger than respondent	19.6
Older than respondent	28.6
Same age group	23.2
Can not say/do not know	28.6
Types of partner	
Girl friend/fiancée	21.5
Relative/married female neighbour	35.7
Stranger/commercial sex worker	33.9
Known male person	08.9
Marital status of the partner	
Never married	35.7
Currently married	32.1
Unknown	32.1
Nature of debut	
Planned	36.1
Unplanned	63.9
Location	
Brothel	03.6
Lonely places	10.7
Own/some body's house	60.7
Hotel/lodge/rented house	25.0
Initiation taken by	
Mutual consent	57.1
Respondent persuaded/forced the partner	12.5
Female partner persuaded/forced	26.8
Male partner persuaded/forced	03.6
Condom use	
Yes	21.4
No	78.6
Alcohol consumption	
Yes	05.4
No	94.6

Surprisingly, one-third of the youth have had their sexual debut with a stranger (an international or domestic tourist) or commercial sex worker putting them at the greater risk of acquiring STI as well as AIDS. The incidences of unprotected sex seem to be at higher side as only 21 per cent of the youth has reportedly used condom in their sexual debut. Although a few youth have reportedly consumed alcohol prior to sex, nearly two-third of sexual debuts were unplanned. '.....*that night I was just thinking of going to my room, she started changing her dress in front of me. I became a bit excited. She noticed my excitement and asked me whether I ever experienced intercourse to which I replied negatively. She then started laughing. I did not say anything as she was almost naked and I was watching what was happening. She came nearer and started kissing me leaving me uncontrolled by that time and we just did it.........there and then' (21 years old travel agent, who accompanied a foreign tourist to Konark')'*.

The location of sexual debut in most cases is either own home or somebody else's home (about 64 per cent cases) followed by a rented place (23 per cent). Although most of the sexual debuts (57 per cent) are with mutual consent, nevertheless, there are about one-third of the instances where it is because of force or persuasion by the partner. '..... *She (foreign tourist) told me to come to her hotel room on the pretext that I take her photographs and hence I accompanied her to her hotel room. In between she indicated her intension to have sex. She offered me Rs. 500/-for the same. Though I was initially not interested in her offer but later agreed because by any standard Rs. 500/-is not a small amount for a photographer like me. I could not control myself.........after some time, the drama (khela) was over and I came out with a hefty sum in my pocket (20 year old, Photographer)'*.

Determinants of Premarital Sex

In the first logistic regression model (Table14.4), age and education besides the life style appear to play crucial role in influencing the likelihood of youth indulging in premarital sex. It is evident that with advancement of age of the youth, the likelihood of premarital sex also increases and the relationship is statistically significant too. In case of education, the likelihood of premarital sex is less among those educated above high school in comparison to their counterparts who are educated up to high school. Results also indicate that more the risky life style, more is the chance of being indulging in premarital sex. Again, the probability of indulgence is more among the youth who are not working than their counterparts who are either in formal or informal sector or students. Further, the odds of engaging in premarital sex is more among youth belonging to general category than those belonging to SC/ST/ OBC but not statistically significant.

Table 14.4: Odds ratios from logistic regression analysis examining the association between selected characteristics and premarital sex

Explanatory Variables	*Model 1 (=156)*	*Model 2 (N=149)*
1	*2*	*3*
Age		
15-20®		
21-24	5.545***	6.22***
Education		
Up to high school®		
Above high school	.241*	.098***
Occupation		
Student®		
Working	1.175	.596
Not working	3.144	2.151
Life style index		
Moderate/no risky®		
High risky	3.135*	2.855*
Caste		
SC/ST/OBC®		
Others	2.743	2.236
Region		
Region-1&3®		
Region-2&4	.419	.461
Type of family		
Nuclear®	—	
Joint	—	1.472
Presence of sister		
No®	—	
Yes	—	.233**

(Contd...)

1	2	3
Family income		
Low®	—	
Medium/high	—	3.854*
Standard of living of the household		
Low/medium®	—	
High	—	.481
Constant	.169	.345

***P< 0.01 **P< 0.05 *P< 0.10.
—Not Included.

Again, after adding type of family, presence of sister, family income and standard of living of the household to the analysis (model 2), the association of age, education, life style and caste of the respondents with premarital sex has remained intact. It is found that the likelihood of premarital sex decreases among the working respondents than the students. Presence of sister comes out as a strong predictor of premarital sex of the youth as the table shows the likelihood of premarital sex significantly decreases with presence of sister in the family. Further, the chances of premarital sex is more among the youth of medium or high income families than their counterparts of low income families. In case of standard of living of the household, the odds of premarital sex among the youth of low or medium standard of living household is more than those belonging to household of higher standard of living. So far as different regions are concerned, it seems that the likelihood of premarital sex is more in region-1& 3 as compared to region-2 & 4.

Sexual Health Problems and Treatment Seeking

Fig. 14.1 clearly exemplifies the sexual health problems youth in Puri have suffered/suffering from during the past three months prior to the survey date. Involuntary loss of semen (Dhatu Padiba) emerges out as the most common sexual problem as about one out of every fifth respondent reportedly is suffering from it. Swapnadosh/ wet dream is another commonly reported sexual problem (19 per cent) followed by 'Kamjori' (sexual weakness). Among sexually active youth 'involuntary loss of semen' is reported as the most common sexual health problems as more than 40 per cent of the youth complained of the same. Some of the other problems reported by the youth are 'Kamjori' (sexual weakness) followed by 'burning during urination' and 'swapnadosh or wet dream'. Here it may be mentioned that almost one-third of the youth in the study area reportedly are suffering from at least one of the sexual health problems in last three months preceding the survey and the health problems are relatively more among the sexually active youth, possibly because of unsafe sexual practices.

It has observed that nearly half the respondents suffering from any sexual health problem do not seek treatment for the same (fig. 14.2). While majority of the youth (59 per cent) have reportedly do not feel the need for the same since they believed it to be normal situation, stigma attached to these diseases prevent a quarter of respondents from treatment seeking. '.... *if people will know that I am suffering from this problem, they will laugh at me and tease me...... (19 year old student)'*. Inclination towards traditional methods of treatment (Ayurveda and Homoeopathy) is what comes out from the analysis as more than half of the respondents who have sought treatment have gone for it. Another important finding is that youth generally prefer to take treatment from the doctors or any other source (preferably ayurvedic/ homoeopathic) usually located away from their place of residence. Fear of being noticed, cheap price and privacy besides quality of care are found to be the reasons. '... *even if you use their medicine, no body will come to know ...(boy aged 18 years)'*.

In addition to the above discussed health problems, findings from qualitative analysis come out with unwanted fatherhood as result of unprotected sex. One out of every four sexually active respondents have revealed that they either have persuaded or forced the partner for induced abortion in order to avoid unwanted fatherhood. ....' *Once she told me that she is going to be the mother of my child. I was a bit worried. I persuaded her to go for abortion (peeta safa). But in the due course her family members came to know about it and to save the prestige of the family, they forced her to go for abortion secretly. But it took less time to come to everybody's notice in the village. I was frightened and left the village for around one year'* (24years migrant *worker*).

Fig. 14.1: Per cent of Respondents, Suffered/Suffering from Self Reported Sexual Health Problems in Last Three Months Preceding the Survey

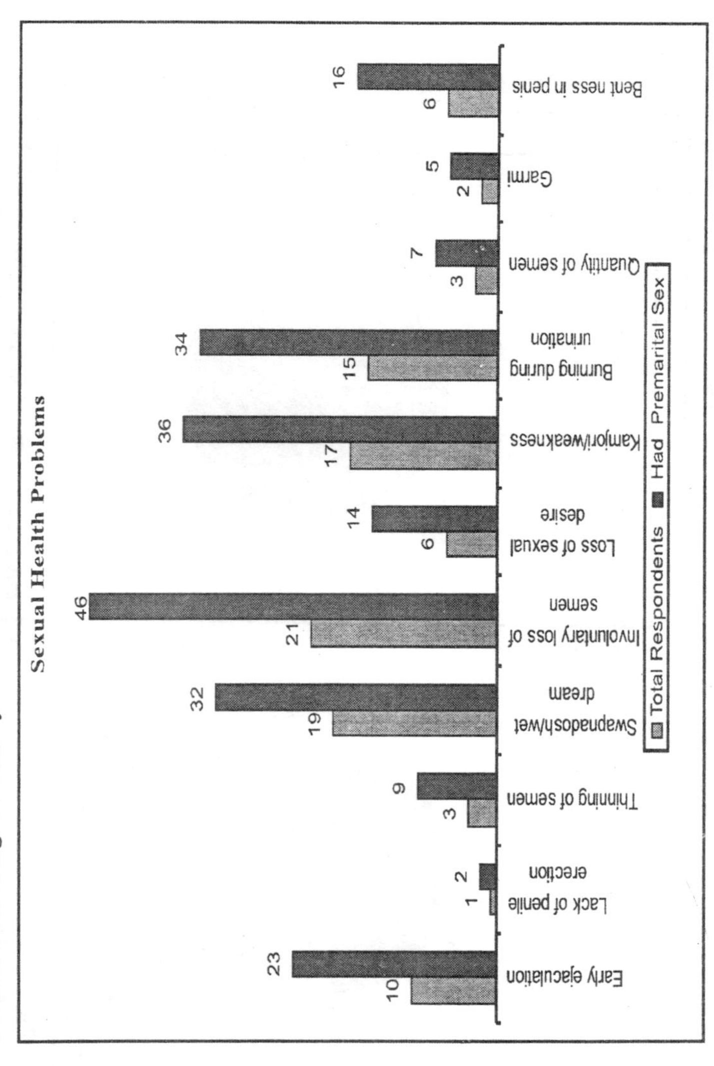

Fig. 14.2: Treatment Seeking Behaviour

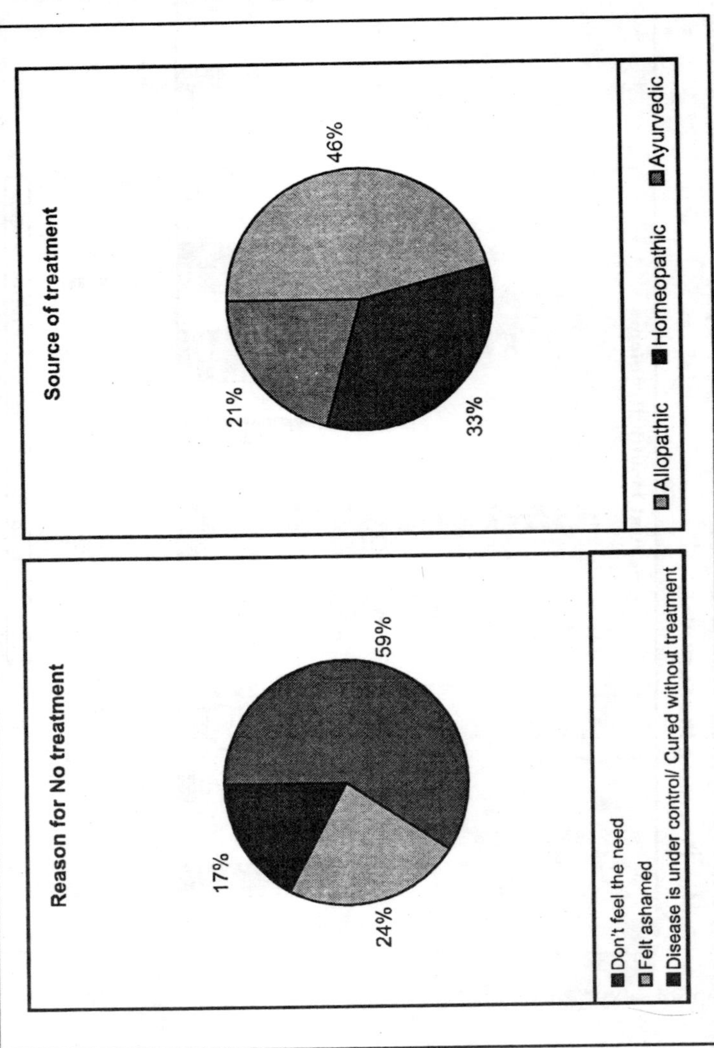

Conclusion

Out of the youth covered in the study, 36 per cent have reported having experienced premarital sex. Further, about half of youth who have made sexual debut also have multiple partners and often a CSW or a stranger. The point, which needs attention, is that most of the sexual debuts are unplanned as well as unsafe and thus increases the risk of STI, unwanted pregnancy, induced abortion and HIV/AIDS. Unwanted fatherhood as a result of unprotected sex is not uncommon and often leads to forced induced abortions and has wider social implications both on the individuals and their families. Increase in the age at marriage, peer pressure and impact of tourists besides some prevalent cultural norms emerge out as the important reason for higher incidence of premarital sex among the youth in Puri.

Qualitative data clearly bring out that involvement of youth in premarital sex with foreign tourists, largely for financial incentives. The present study reveals that the youth staying away from the family or staying with a friend are more likely to involve in premarital sex. Besides this, it has found that presence of sister in the family is negatively associated with premarital sex of the youth. Further, it may be brought to the notice of the reader that those youth, who are either illiterate or educated up to middle the sexual involvement is more than 50 per cent, which declines with increase in literacy level. Above all, almost a third of the respondents are reportedly suffering from at least one of the sexually transmitted infections and nearly half of them do not seek treatment due to stigma attached to these diseases and/or failing to understand the severity of the diseases.

The dynamic role of the youth in nation building, in preserving the social and cultural values of the nation is important in all ages. The present research suggests the need for condom promotion as a barrier against both STI and unwanted pregnancy. Besides this, programmes promoting abstinence from risky sex needs constant emphasis with increasing focus on those revealing deviant sexual behaviour.

REFERENCES

Abraham, Leena and Anilkumar, K, 1999, Sexual Experiences and Their Correlates Among College Students in Mumbai City, India. *International Family Planning Perspectives,* 25(3): 139-146.

Abraham, Leena, 2002, *Bhai-behen,* True Love, Time Pass: Friendships and Sexual Partnerships Among Youth in an Indian Metropolis. *Culture, Health and Sexuality,* 4(3): 337-353.

Advocates for Youth, 1978, *Serving the Future: An Update on Adolescent Reproductive and Sexual Health Programmes in Developing Countries.* 2nd Edition, Washington DC.

Asian Institute of Marketing Studies (AIMS) Research Group, 1998, Men's' Sexual Health Concerns: A Free Listing Approach. Unpublished Draft Paper, Bhubaneswar.

Brewster, KL., Cooksey, EC., Guilkey, DK and Rindfuss, RR, 1998, The Changing Impact of Religion on the Sexual and Contraceptive Behaviour of Adolescent Women in the United States. *Journal of Marriage and the Family,* 60(2): 493-504.

Djamba, YK, 1997, Financial Capital and Premarital Sexual Activity in Africa: The Case of Zambia. *Population Research and Policy Review,* 16(3): 243-257.

GoI, 1997, *National Youth Policy–1997* (Draft). Ministry of Youth Affairs and Sports, Government of India.

GoI, 2000, *National Population Policy–2000.* Ministry of Health and Family Welfare, Government of India.

GoI, 2002, *National Health Policy–2002.* Ministry of Health and Family Welfare, Government of India.

Goparaju, L, 1994, Discourse and Practice: Rural Urban Differences in Male Students' Sexual Behaviour in India. Paper presented for IUSSP Seminar on Sexual Subcultures, Migration and AIDS, Thailand.

Hill, ZE., Cleland, JG and Ali, MM, 2004, Religious Affiliation and Extramarital Sex in Brazil. *International Family Planning Perspectives,* 30(1): 20-26.

ICPD, 1994, *Newsletter of the International Conference on Population and Development,* Cairo, Number 15, New York.

Jejeebhoy, S and Sebastian, M., 2003, *Actions that Protect: Promoting Sexual and Reproductive Health and Choice Among Young People in India.* Population Council, New Delhi.

Jejeebhoy, S and Sebastian, M., 2004, Young Peoples Sexual and Reproductive Health. In: S. Jejeebhoy (eds.), *Looking Back and Looking Forward,* Rawat Publication, Jaipur.

Mamdani, M., 1999, Adolescent Reproductive Health Exposure of Community-based Programme. In S. Pachauri (eds.), *Implementing a Reproductive Health Agenda in India: The Beginning*, Population Council, New Delhi.

Miller, BC., Benson, B and Galbraith, KA, 2001, Family Relationships and Adolescent Pregnancy Risk: A Research Synopsis. *Developmental Review*, 21:1-38.

Miller, KS and Whitaker, DJ, 2001, Predicators of Mother-Adolescent Discussions About Condoms; Implications for Providers Who Serve Youth. *Paediatrics*, 108(2): 28.

National AIDS Control Organization, 1994, *Country Scenario: An Update*, Govt. of India. Ministry of Health and Family Welfare, New Delhi.

Radhakrishna, A., Gringle, RE and Greenslade, FC., 1997, Identifying the Intersection: Adolescent Unwanted Pregnancy, HIV/AIDS and Unsafe Abortion. Carrboro NC: IPAS.

Ramasubban, R., 1992, Sexual Behaviour and Condition of Health Care: Potential risks for HIV Transmission in India. In T. Dyson (ed.), *Sexual Behaviour and Networking: Anthropological and Socio-cultural Studies on the Transmission of HIV*, IUSSP, Liege.

Registrar General of India, Final Population Total, New Delhi, 2001.

Singh, S, 1997, Men, Misinformation and HIV/AIDS in India, Toward a New Partnership: Encouraging the Positive Involvement of Men as Supportive Partners in Reproductive Health. Population Council, New York.

United Nations, 2001, *World Population Prospects: The 2000 Revision. Volume II, The Sex and Age Distribution of the World Population*, New York.

United Nations, 2003, *Promoting the Millennium Development Goals in Asia and the Pacific: Meeting the Challenges of Poverty Reduction*, New York.

Varga, CA, 1999, South African Young Peoples' Sexual Dynamics: Implications for Behavioural Responses to HIV/AIDS. In J.C. Caldwell, P. Caldwell, J. Anarfi, K. Awusabo-Asare, J. Ntozi, I. O. Orubuloye, J. Marck, W. Cosford, R. Colombo and E.Hollings (Eds.), *Resistances to Behavioural Change to Reduce HIV/AIDS infection*, Canberra: Health Transition Centre, The Australian National University.

Verma, R., Mahendra, VS., Pulerwitz, J., Barker, G., Vandam, J and Flessenkaemper, S,2004, From Research to Action: Addressing Masculinity and Gender Norms. *Indian Journal of Social Work*, 65(4): 634-654.

Index

H

I

J

K

L

M

N

O

P

Q

R

S